Stereoselectivity in Organic Synthesis

Garry Procter

Professor of Organic Chemistry, University of Salford

OXFORD NEW YORK TOKYO
OXFORD UNIVERSITY PRESS
1998

Oxford University Press, Great Clarendon Street, Oxford OX2 6DP

Oxford New York
Athens Auckland Bangkok Bogota Bombay Buenos Aires
Calcutta Cape Town Dar es Salaam Delhi Florence Hong Kong Istanbul
Karachi Kuala Lumpur Madras Madrid Melbourne Mexico City
Nairobi Paris Singapore Taipei Tokyo Toronto Warsaw

and associated companies in
Berlin Ibadan

Oxford is a trade mark of Oxford University Press

Published in the United States
by Oxford University Press Inc., New York

A catalogue record for this book is available from the British Library

Library of Congress Cataloging in Publication Data
(Data available)
ISBN 0 19 855957 7

Typeset by the author

Printed in Great Britain by Bath Press Ltd., Bath

Series Editor's Foreword

Synthesis is the central arena of modern organic chemistry providing specifically designed and precisely constructed materials on which many other sciences rely. The current high level of precision demanded of synthesis, including the provision of single enantiomers as well as single diastereoisomers, makes the understanding of the factors which allow control of the stereo- selectivity in organic synthesis an essential and fundamental topic for all students of chemistry.

Oxford Chemistry Primers have been designed to provide concise introductions relevant to all students of chemistry and contain only the essential material that would be covered in an 8–10 lecture course. The present primer by Garry Procter presents the basic concepts of stereo- selective organic synthesis in a very logical and readily understandable fashion. This primer will make essential reading for apprentice and master chemist alike.

Professor Stephen G. Davies
The Dyson Perrins Laboratory
University of Oxford

Preface

Organic chemistry is a vast and important subject. Within organic chemistry, organic synthesis is a discipline which occupies a place of special importance. At the most basic level it allows for the preparation of compounds and materials which have become essential for the way we live. At the other extreme, it can provide the opportunity to invent, and to synthesize molecules which are new to the world, and which could possess valuable properties.

Organic synthesis can also make demands at the highest scientific level, and at the extreme edge of the subject, advances require and inspire creativity equal to that of any other discipline. Planning and executing organic synthesis is both an art and a science which calls upon many human abilities, from the abstract design and planning stages, to the practical skills required to make the plan succeed. The three-dimensional structures of organic molecules are often crucial in determining their properties, and their complexity and beauty serve to inspire organic chemists from one generation to another.

Such three dimensional structures require efficient methods for their stereoselective synthesis. The methods of stereoselective organic synthesis are important at all stages of this type of synthetic endeavour. Planning has to take account of what is possible, or what might be possible to achieve. Methods for stereoselective synthesis should not therefore be seen in isolation, as an end in themselves. In this area, application is all.

The best stereoselective synthetic methodology should allow for very high levels of three- dimensional control, but at the same time be easy and reliable in the laboratory. In this introduction I have tried to include some of the best examples of such methodology, and to use simple (often simplistic!) theory to account for the observed stereochemical aspects in these reactions.

Any good organic textbook would complement this Primer, and a set of basic molecular models will be invaluable. In addition, I particularly recommend A. J. Kirby, Stereoelectronic Effects Oxford Chemistry Primer no. 36, Oxford, 1996, as an excellent introduction to this aspect of organic chemistry, so important in understanding stereoselective organic synthetic methods.

I agree with Professor Kirby that 'learning organic chemistry is hard work'. But once the threads which run through organic chemistry are understood, it becomes extremely rewarding and intellectually fulfilling. Nowhere more than in the area of stereoselective synthesis, where the application of existing methods to taxing target structures, often provided by Nature, and the invention of new synthetic methods, provide a never ending series of challenges to the creativity of organic chemists. If this Primer helps you in setting out on the adventure that is organic synthesis, then I will be happy.

Salford
1998

G. P.

Contents

1 Introduction

1.1 General considerations

Why worry about stereochemistry?

Many properties of organic compounds are closely related to their precise three-dimensional structure. At a very basic level, an organic molecule can be viewed as an array of connected atoms, held together by bonding electrons, which presents to the 'outside world' a surface of varying electron density. In this simplistic picture, the organic molecule could be thought to interact with the 'outside world' by interaction of this surface with its surroundings or with other molecules. Inevitably, this surface extends into three dimensions, and this simple fact, that the molecules of an organic compound possess a defined three-dimensional structure, has profound consequences. These might include the rate of its reactions, its physical properties, and its biological properties such as taste, smell, and efficacy as a pharmaceutical. Given the importance of the three dimensional structure of organic compounds, and that all organic compounds arise as a result of a previous reaction or reaction sequence, it is not surprising that an area of chemistry which encompasses organic reactions and stereochemistry is central to organic chemistry. It is this area which is the subject of this introduction to stereoselectivity in organic synthesis.

About this book

This book is intended to provide an introduction to the area of stereoselective organic reactions, and to highlight their usefulness in organic synthesis where appropriate. In terms of stereochemistry, we will start with the basics, and also address the inevitable jargon which is associated with the topic. This jargon will be kept to a minimum. As for the reactions which are covered, we will look at the basic processes which are used to construct organic compounds. Of course, there is a dauntingly wide array of such reactions, but we will concentrate on commonly encountered fundamental types of reactions. Particular emphasis will be placed on carbonyl additions, enolate functionalization, aldol reactions, additions to C–C double bonds, reduction, oxidation, rearrangements, and some enzyme catalysed reactions. Before considering these individual reaction types, an overview of the various approaches to stereoselective organic synthesis will be presented, and the relationships between them explored.

The principles which underlie stereoselective and stereospecific organic reactions, orbitals and their interactions, transition state structures, and mechanistic considerations will be treated together. Subsequently these principles will be applied when appropriate to the individual types of reaction.

1.2 Stereochemical considerations

Grignard reagents, such as MeMgBr, react as carbon nucleophiles, in this case as Me^-.

The term 'chiral' is derived from the Greek for hand '*cheir*', and was first used in 1884.

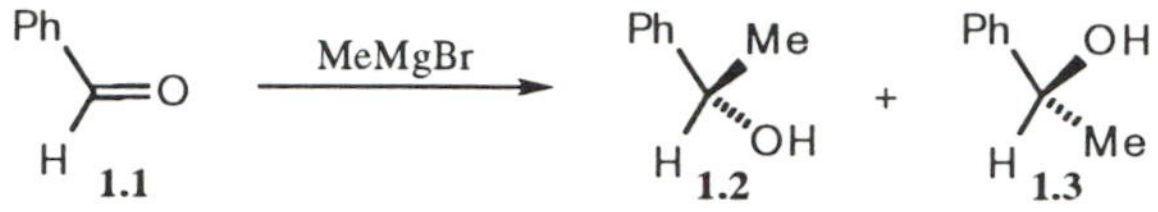

For most of the time, we will be concerned with reactions which involve the formation of tetrahedral, or sp^3, carbon atoms within a molecular framework. Such reactions are widespread. The addition of a Grignard reagent to an aldehyde **1.1** is just such a reaction. When the nucleophilic group is different to the substituent attached to the carbonyl carbon, then two products **1.2** and **1.3** are formed which are enantiomers (Fig. 1.1). The products are chiral.

Ph, H, O **1.1** —MeMgBr→ Ph, Me, OH, H **1.2** + Ph, OH, Me, H **1.3**

Fig. 1.1 Formation of a stereocentre from a trigonal carbon

Absolute configuration; the spatial arrangement of atoms in a chiral molecule, described as either *R* or *S*.

The Latin words for right (*rectus*) and left (*sinister*) give us the descriptors *R* and *S* respectively.

The carbon atom which is 'responsible' for the chirality is often referred to as a chiral centre or stereocentre. Enantiomers have opposite absolute configurations, and these are given the descriptors *R* and *S*. The assignment of a stereocentre as either *R* or *S* follows from the Cahn-Ingold-Prelog (CIP) convention. Details of this convention can be found in any major undergraduate text on organic chemistry.

In the nucleophilic addition of **1.1** which produces **1.2** and **1.3** (Fig. 1.1) the enantiomers are formed in equal quantities, giving a racemic mixture. Enantiomer **1.2** would arise by addition from above the plane of the drawing, and **1.3** by addition from below. Stated another way, the two enantiomers arise by addition from the two faces of the carbonyl group. In order to avoid confusion as to which face is which ('upper' and 'lower' face will not do, as this depends on how the structure is drawn). The faces can be assigned a descriptor using a procedure similar to that which is used to assign absolute configurations (*R* and *S*).

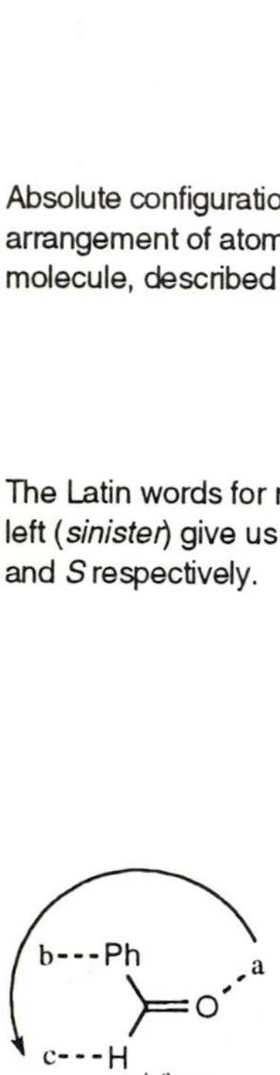

b---Ph, a, O, c---H
si face
1.1

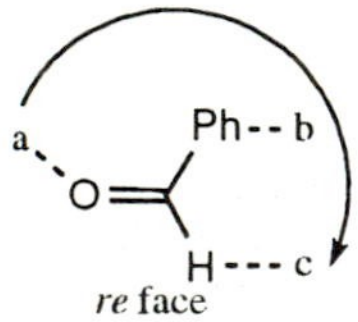

a, O, Ph--b, H---c
re face
1.1

The carbonyl group is drawn in the plane of the paper, and the three groups attached are assigned a priority a, b, and c, using the same rules as for the assignment of absolute configuration. If the sequence a, b, c is anticlockwise it is the *si* face, and if it is clockwise it is the *re* face.

When the nucleophile attacks from the *si* face, enantiomer **1.2** is formed, whereas enantiomer **1.3** is formed when reaction takes place from the *re* face. These reactions take place at identical rates, because the transition states are equal in energy by virtue of them being enantiomers (Fig. 1.2).

The mirror plane relationship between the products (**1.2** and **1.3**) is retained in the transition states, and in the reaction profiles. This means that the transition state energies must be equal by symmetry, and the reactions which produce **1.2** and **1.3** must be exactly equal in rate. It follows that an equal amount of **1.2** and **1.3**, a racemate, must be obtained.

When addition to opposite faces of a planar group produces enantiomers in this way, the faces are said to be enantiotopic. Atoms or groups within a molecule can also be enantiotopic. Consider the diester **1.4** (Fig. 1.3). The two ester groups are enantiotopic because reaction at one (CO_2Me_X) gives enantiomer **1.5** whereas reaction at the other (Y) gives enantiomer **1.6**.

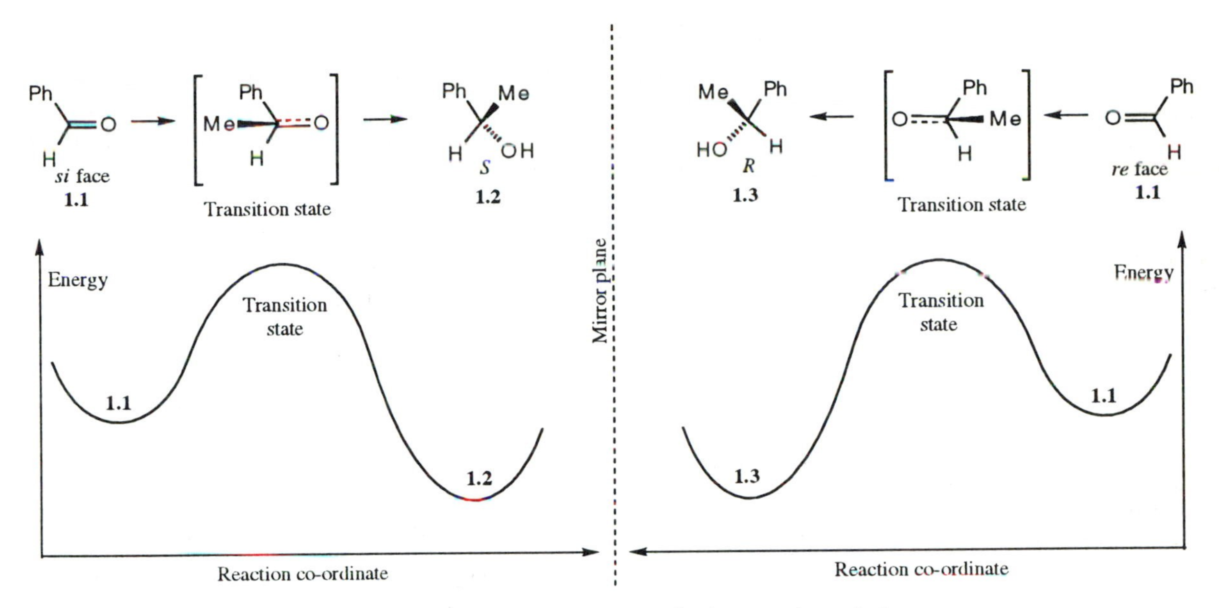

Fig. 1.2 Enantiomeric transition states lead to racemic products

Enantiotopic ligands can be assigned descriptors, *pro-R* and *pro-S*, using a modification of the normal sequence rules. The groups attached to the stereocentre are assigned priorities as usual. By definition, two of these will be identical (the two ester groups in **1.4**). To assign the descriptor to one of these two groups, the group in question is arbitrarily assigned the higher priority. If this results in the *R* configuration, the group is said to be *pro-R*, and *pro-S* if it results in the *S* configuration. This process is outlined in Fig. 1.4 for diester **1.4**.

Hydrolysis of ester group X → **1.5** (*R*) | Mirror plane | **1.6** (*S*) ← Hydrolysis of ester group Y

Fig. 1.3 Consequences of reactions at enantiotopic groups

So far the discussion has been restricted to compounds which are prochiral, and reactions which give products containing only one stereocentre. When we consider reactions which involve a starting material which contains a

A carbon which possesses enantiotopic faces or enantiotopic groups is said to be **prochiral**.

For CO_2Me_X:– Me = d; Bu^tO = c; CO_2Me_Y = b; CO_2Me_X = a

a → b → c anticlockwise – CO_2Me_X is *pro-S*

For CO_2Me_Y:– Me = d; Bu^tO = c; CO_2Me_X = b; CO_2Me_Y = a

a → b → c clockwise – CO_2Me_Y is *pro-R*

Fig. 1.4 Stereochemical descriptors for enantiotopic groups

stereocentre (often referred to as a pre-existing stereocentre), and which result in formation of a new stereocentre, the situation is fundamentally different. A typical example of this type of reaction is shown in Fig. 1.5.

A single (*S*) enantiomer of ketone **1.7** reacts with phenylmagnesium iodide. As expected, reaction from the *re* and *si* faces of the carbonyl group gives products which are stereoisomeric. Unlike the previous reaction (Figs. 1.1 and 1.3) the two products are not enantiomers, as the original stereocentre is unchanged in the reaction and must still be *S*. In the enantiomer of **1.8** both stereocentres would possess the *R* configuration. The products **1.8** and **1.9** are diastereoisomers.

Diastereoisomers; stereoisomers not related as mirror images.

PhMgI

1.7

1.8 *re* face addition

1.9 *si* face addition

Fig. 1.5 Formation of diastereoisomers on addition to a chiral ketone

Addition to the two faces of **1.7** gives diastereoisomers. Such faces are referred to as being diastereotopic, as are substituents which on replacement would also give diastereoisomers. For example, the hydrogen atoms of the CH_2 group in **1.5** (Fig. 1.3) are diastereotopic. The diastereotopic relationships in **1.5** and **1.7** are illustrated in Fig. 1.6.

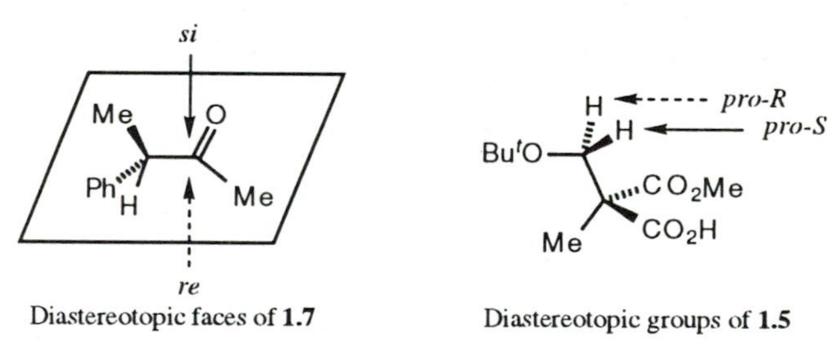

Diastereotopic faces of **1.7**

Diastereotopic groups of **1.5**

Fig. 1.6 Diastereotopic groups and faces

1.3 Diastereoselective reactions

See Section 1.5 for a discussion of stereoselective and stereospecific reactions.

Any reaction which could produce two (or more) possible stereoisomeric products is capable of producing one in excess. The most straightforward sort of reaction of this type is one in which a new stereocentre is created in a molecule which already contains (at least) one stereocentre. We have already met such a reaction in Fig. 1.5. This reaction will be used to illustrate some of the important aspects of this type of reaction.

Do not confuse the terms stereoselective and stereospecific. They have quite different meanings. See Section 1.5.

The relative amounts of the diastereoisomeric products **1.8** and **1.9** will be determined by the difference in energy between the two transition states. As the products are diastereoisomeric, the transition states which lead to each of the products are themselves diastereoisomeric. This means that there is no reason for them to be of equal energy. In general, if transition states are diastereoisomeric for any reason then they can be of different energies, and the

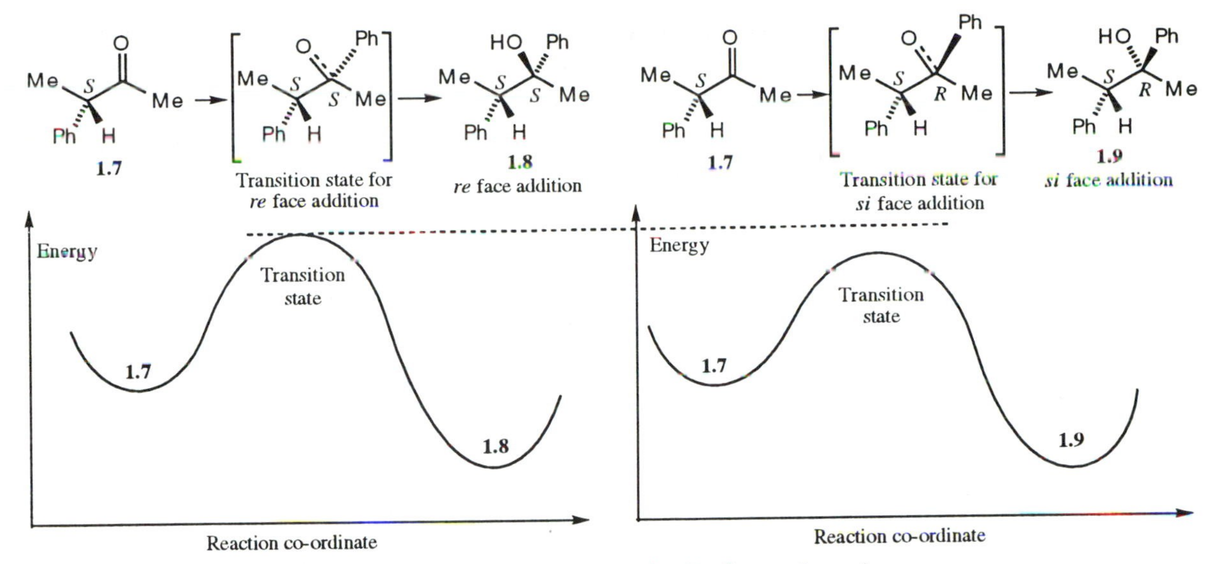

Fig. 1.7 Diastereoisomeric transition states lead to diastereoisomeric excesses

stereoisomeric products would then be formed in different amounts. This is illustrated by Fig. 1.7. In this reaction of **1.7** addition from the *si* face has a lower transition state energy and the major product is **1.9**. Assuming that the reaction is effectively irreversible then the ratio of the two products is related directly to this difference in transition state energies. In the reaction shown in Fig. 1.7 this leads to **1.9** and **1.8** being formed in a ratio of 91.5:8.5. Reactions such as this, in which two or more diastereoisomeric products are possible, and in which the products are formed in unequal amounts, are described as stereoselective or diastereoselective. The extent of the diastereoselectivity of a reaction is sometimes described simply as the ratio of the products (as above), but more usually as the diastereoisomeric excess (abbreviated to d.e.), which is simply the excess of one diastereoisomeric product over the other.

For the reaction of **1.7** (Fig. 1.7) the d.e. is 83 per cent (91.5%–8.5%).

Generalizing from the preceding discussion, whenever transition states are diastereoisomeric, we can expect to observe the selective formation of one stereoisomer assuming the reaction is irreversible (kinetic control). Whenever a stereocentre is being formed in a reaction, and another stereocentre is present in the transition state, the transition states are diastereoisomeric.

This 'other' stereocentre could be present in the substrate (as Fig. 1.7), a reagent, or a catalyst.

In the case of the reaction of **1.7** discussed above, with a pre-existing stereocentre, the stereoselectivity is sometimes referred to as 'internal stereocontrol' or 'simple diastereoselectivity'. In effect the diastereoselectivity arises because the substrate (starting material) is chiral. This is not the only way in which transition states can be rendered diastereoisomeric. In addition to the substrate, there might be a reagent, catalyst, or solvent molecules involved in the transition state. Should any of these be chiral, then stereoselective reactions are possible. We will now consider the consequences of this in Section 1.4.

1.4 Enantioselective reactions

By 'normal' we mean that there is no chirality associated with the transition state.

As discussed in Section 1.2, a 'normal' reaction which gives enantiomeric products is required to produce each in equal amounts because of the enantiomeric relationship between the two transition states. If a component of the transition state other than the substrate is chiral, and present as a single enantiomer itself, then the enantiomeric products need no longer be formed in equal amounts. One of the enantiomers should be formed in excess. This is known as asymmetric synthesis, and the reaction is said to be enantioselective. There are several general methods for achieving asymmetric synthesis, and we will consider the principles behind them in this section.

Chiral reagents

If a reagent which gives enantiomeric products is chiral, and present as a single enantiomer, then it is possible for the reaction to be enantioselective.

1.10 + **1.11** → (*R*)-**1.12** *si* face addition (>99%) + (*S*)-**1.12** *re* face addition (<1%)

Fig. 1.8 A chiral reagent which leads to an enantioselective reaction

The chiral reagent does not have to be used as a single enantiomer, as long as the racemate is not used we could expect some level of enantioselectivity.

Consider the reaction of ketone **1.10** (Fig. 1.8) with the chiral borane **1.11**. The transition states for the formation of the enantiomers of **1.12** are diastereoisomeric, and the reaction now produces the enantiomers in unequal quantities. In this case, the *R* enantiomer predominates. The extent of the selectivity is expressed as the enantiomeric excess (e.e.), and is the percentage of the major enantiomer minus that of the minor ($R(\%) - S(\%)$).

For reaction of **1.10** with **1.11** the e.e. is >99 per cent.

Chiral catalysts

A single enantiomer of the catalyst is not always required. In some cases a catalyst of relatively low e.e. can be almost as effective as one of 100 per cent e.e. In other cases, a catalyst of low e.e. will simply give a product of correspondingly low e.e.

A catalyst, by definition, accelerates a reaction. This means that it must be involved in the transition state. It follows that if a particular catalyst can be made chiral, and a single enantiomer is used, then it should allow enantioselective reactions. A schematic representation of such a catalytic process is presented in Fig. 1.9. A catalyst–substrate complex is formed initially. The reaction in which the stereocentre is created then takes place, under the influence of the chiral catalyst. This results in a catalyst–product complex, which then dissociates to product and catalyst. The catalyst is then free to undergo another cycle.

The transition state for the reaction, somewhere between the catalyst–

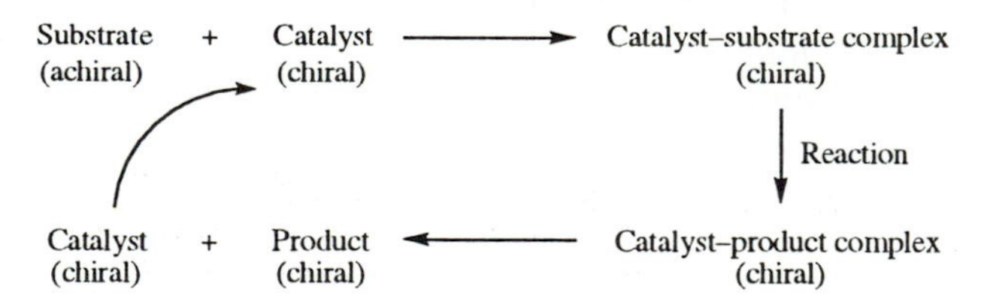

Fig. 1.9 Schematic representation of enantioselective catalysis

1.13 (2%)

(*S*)-**1.14** e.e. = 99% (*R*)-**1.14**

Addition to the *si* face (99.5%) Addition to the *re* face (0.5%)

Fig. 1.10 Asymmetric catalysis of nucleophilic addition to a C=O group

substrate and the catalyst–product complex, must involve the single enantiomer of the catalyst and the 'developing' product stereocentre. By analogy with the discussion above concerning enantioselective reactions of chiral reagents, this should result in one possible product enantiomer forming in excess. This process is known as asymmetric or enantioselective catalysis.

A typical example of such asymmetric catalysis, which again involves addition to an aldehyde is shown in Fig. 1.10. Dialkylzinc reagents react very slowly with aldehydes, but catalytic amounts of 1,2-aminoalcohols will accelerate addition to the carbonyl. If this 1,2-aminoalcohol is chiral and used as a single enantiomer (**1.13**) then the reaction is enantioselective.

As can be seen from the reaction shown in Fig. 1.10, asymmetric catalysis can be an extraordinarily effective approach to enantioselective synthesis. Only 2 per cent of the chiral catalyst is needed to produce the alcohol **1.14** with an e.e. of 99 per cent. Moreover, as the catalyst is unchanged after reaction is complete, it could even be recovered and used again. However, there are relatively few types of reactions which are susceptible to such highly enantioselective catalysis.

An alternative approach, which takes advantage of diastereoselective reactions (Section 1.3) and the sequence of events which is involved in asymmetric catalysis (Fig. 1.9). This approach to asymmetric synthesis, the use of chiral auxiliaries, is outlined below.

Chiral auxiliaries

A chiral auxiliary is a compound which is attached to a prochiral starting material, thereby causing groups or faces which were enantiotopic to become diastereotopic (Section 1.2). A general representation of the overall process is provided in Fig. 1.11.

While the overall scheme for using chiral auxiliaries resembles that for asymmetric catalysis (Fig. 1.9), there are some important differences. The

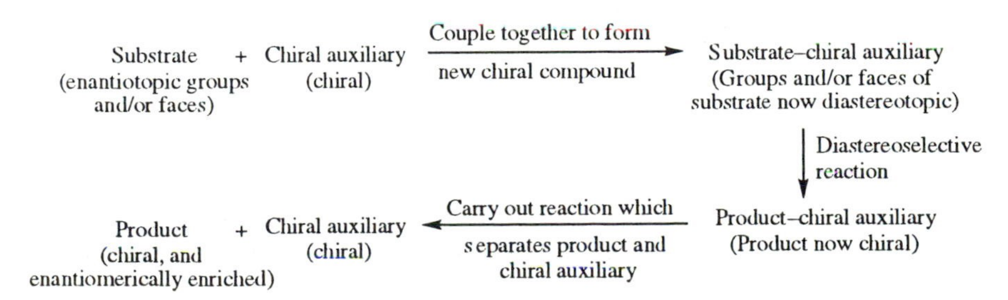

Fig. 1.11 Schematic representation of use of achiral auxiliary

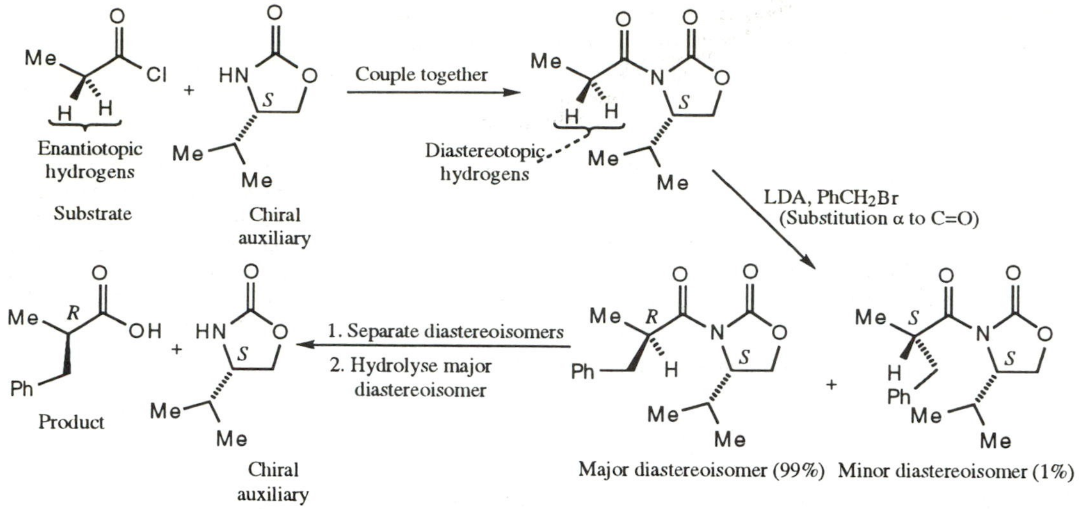

Fig. 1.12 Use of a chiral auxiliary to control substitution α to a C=O group

catalysis process is dynamic, with the catalyst–substrate and catalyst–product complexes being intermediates in the reaction. When a chiral auxiliary is used the process is no longer dynamic.

In the first step, the chiral auxiliary is attached to the substrate to produce a new, discreet, and isolated compound (not an intermediate). The reaction is then carried out, and the product diastereoisomers may be separated and purified. Each of these is then subjected to a reaction which separates the chiral auxiliary and the desired product. An example of this approach is presented in Fig. 1.12.

At first sight the use of a chiral auxiliary might seem rather clumsy when compared to an asymmetric catalytic process, and undoubtedly this is true, provided that an efficient catalytic process exists. If a highly enantioselective catalytic reaction is not available, then use of a chiral auxiliary is often the method of choice.

Moreover, asymmetric catalysis is *enantioselective*, which means that the product is a mixture of enantiomers. If the e.e. is not very high then the minor enantiomer must be separated from the major, and enantiomer separation can be very difficult. If a chiral auxiliary is used, the process is *diastereoselective*, and as the products are diastereoisomers, separation is usually easily achieved by 'normal' purification techniques such as column chromatography. Providing that the reaction yield is high and the diastereoisomer separation is easy, a chiral auxiliary which provides only a modest level of diastereoselectivity can be useful in practice.

1.5 Stereospecific and stereoselective reactions

The terms 'stereospecific' and 'stereoselective' are often misused, and should not be confused. Although we have already encountered diastereoselective and enantioselective reactions (Sections 1.4 and 1.5), we will consider these terms again here in the context of selectivity and specificity.

In a stereospecific reaction, the mechanism *requires* a specific product stereochemistry from a particular stereochemistry of the starting material. This implies that the reaction takes place by a mechanism which has a strict stereochemical requirement. In a stereospecific reaction there can be only one possible product stereoisomer, there can be no selectivity.

A typical example is substitution by the S_N2 mechanism, which is discussed in some detail in Chapter 2 (Section 2.3). Substitutions which take place by this mechanism must involve inversion of the configuration of the carbon atom which is reacting. A typical example of this is shown in Fig. 1.13. The mesylate (±)-**1.15** undergoes S_N2 displacement to give *only* (±)-**1.16**. The diastereoisomer (±)-**1.17** cannot be formed if the reaction is an S_N2 process, as this would correspond to *retention* of configuration.

Note that racemic compounds are involved here. Each enantiomer of **1.15** reacts stereospecifically to give the corresponding enantiomer of **1.16**. OMs; The methanesulfonate, or mesylate group (OSO_2Me). Mesylate ($MeSO_3^-$) is a good leaving group, being the anion of a strong acid.

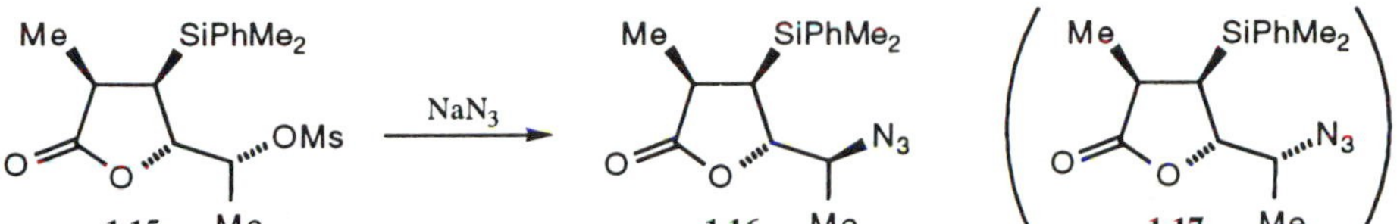

Me SiPhMe$_2$ OMs O O Me **1.15** NaN_3 → Me SiPhMe$_2$ N_3 O O Me **1.16** (Me SiPhMe$_2$ N_3 O O Me **1.17**)

Cannot be formed by S_N2 mechanism from **1.15**

Fig. 1.13 A stereospecific S_N2 reaction

Reactions which take place by the S_N2 mechanism can also be said to be *enantiospecific*. An enantiospecific reaction is one in which the mechanism *requires* that a chiral substrate of a given enantiomeric purity reacts to give a product with the same enantiomeric purity. A simple example is the reaction of (*R*)-2-bromooctane **1.18** with sodium hydroxide (Fig. 1.14), the substitution follows the S_N2 mechanism and the only substitution product which can be formed by this mechanism is (*S*)-2-octanol **1.19**.

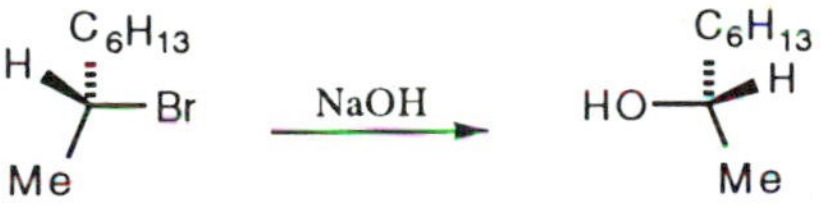

C_6H_{13} H Br Me **1.18** NaOH → C_6H_{13} HO H Me **1.19**

Fig. 1.14 An enantiospecific S_N2 reaction

Formation of (*R*)-2-octanol would require the presence of (*S*)-2-bromooctane.

The stereospecific S_N2 reactions described above involve a starting material which is chiral. Stereospecific reactions are also possible with achiral compounds, a common type being stereospecific reactions of achiral alkenes. In all cases, as required by the definition of a stereospecific reaction, the mechanism requires that a given stereoisomer of the alkene gives rise to a particular diastereoisomer of the product. When the product is chiral, it is formed as a racemate (assuming the absence of a chiral reagent or catalyst), as the faces of the alkene are *enantiotopic* (see Section 1.2). Some typical examples are provided in Fig. 1.15.

In contrast to the types of reactions discussed above, for a *stereoselective* reaction, the mechanism does not prevent the formation of two or more

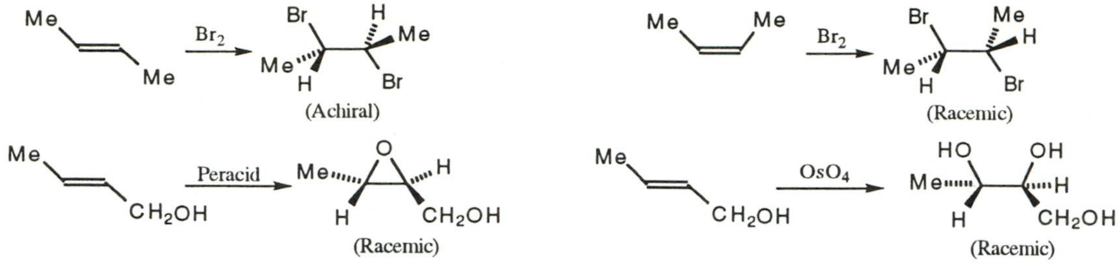

Fig. 1.15 Examples of stereospecific reactions

stereoisomeric products. Three common types of stereoselective reaction are listed below.

(a) Reactions which involve the formation of one or more new stereocentres in a chiral substrate.

(b) Reactions in which two prochiral substrates react such that one new stereocentre originates from each, so that two new stereocentres are formed.

(c) Reactions in which a prochiral substrate reacts with an enantiomerically enriched reagent or catalyst to give a chiral product which is also enantiomerically enriched.

A typical example of a reaction in category (a) is nucleophilic addition to a chiral ketone or aldehyde (Fig. 1.16). As discussed earlier (Section 1.3), one of the possible diastereoisomeric products is often formed in considerable excess, but this selectivity depends on the kinetics of the reaction, rather than the mechanism. In the example used in Fig. 1.7, reaction of **1.7** with phenyl magnesium bromide gives **1.9** with a d.e. of 83%. The mechanism does not dictate which face of the prochiral carbonyl group reacts, it is the difference in *rate* of addition to each face which results in the observed diastereoselectivity.

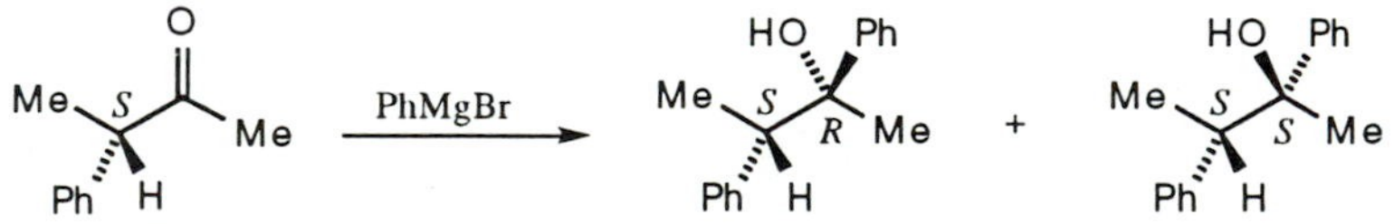

Mechanism of the nucleophilic addition does not preclude the formaton of either **1.8** or **1.9**. Both are formed, with **1.9** predominating (d.e. 83%).

Fig. 1.16 Formation of one new stereocentres in a chiral substrate

Reactions in category (b), in which two prochiral substrates react to give a product containing one new stereocentre from each, are capable of producing two diastereoisomeric products (both racemic if no chiral reagent or catalyst is used). It might be thought that this is unduly complicated, but at least two very important reactions used in organic synthesis fall into this category, aldol reactions (Chapter 5) and Diels-Alder cycloadditions (Chapter 6).

In a typical case, outlined in Fig. 1.17 for an aldol reaction, both reactants are prochiral. Each possesses enantiotopic faces (sp^2) from which the two new stereocentres originate (sp^3), and reactions involving all combination of

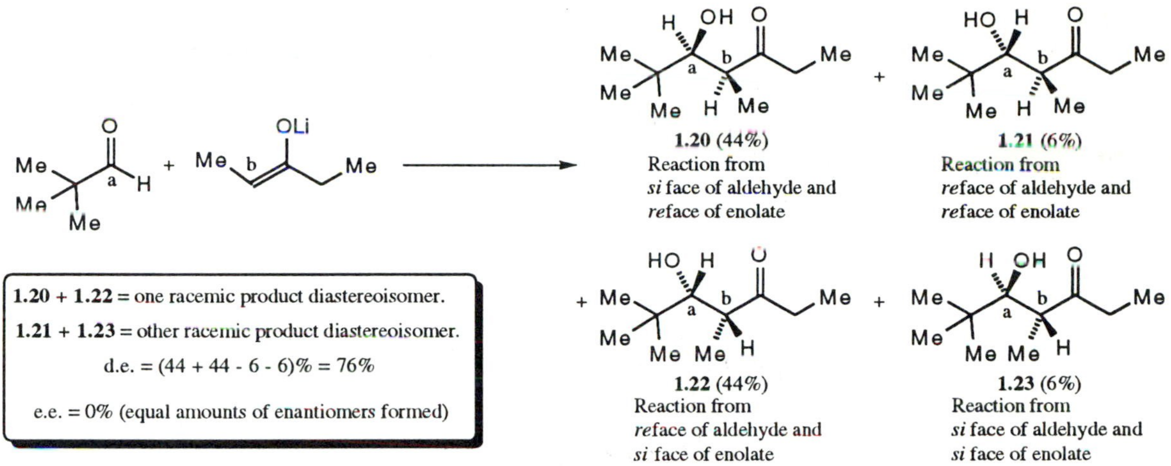

1.20 + 1.22 = one racemic product diastereoisomer.
1.21 + 1.23 = other racemic product diastereoisomer.
d.e. = (44 + 44 - 6 - 6)% = 76%
e.e. = 0% (equal amounts of enantiomers formed)

1.20 (44%) Reaction from *si* face of aldehyde and *re* face of enolate

1.21 (6%) Reaction from *re* face of aldehyde and *re* face of enolate

1.22 (44%) Reaction from *re* face of aldehyde and *si* face of enolate

1.23 (6%) Reaction from *si* face of aldehyde and *si* face of enolate

Fig. 1.17 Two prochiral substrates react to produce two new stereocentres

the faces are possible. This results in the possible formation of two racemic diastereoisomeric products in different amounts. The d.e. of the reaction is 76 per cent, but the e.e. of the two diastereoisomers must be zero.

Enantioselective reactions, corresponding to category (c), in which a chiral reagent or catalyst reacts with a prochiral substrate to produce a non-racemic chiral product, have been discussed at some length earlier in this chapter (see Section 1.4). Example which illustrate the relevant points can be found in Fig. 1.8 and Fig. 1.10.

As you progress through this book you will meet numerous examples of these types of stereoselective reaction. Before we consider the individual reactions in detail, some important general features of reactivity and bond formation need to be covered. These are the substance of Chapter 2.

1.6 Problems

1.1 Enzymes are natural asymmetric catalysts, and are found as single enantiomers. Given this, and the fact that the enzyme pig liver esterase catalyses the hydrolysis of esters, would you expect the hydrolysis of the diester **1.4** to give a mono-ester with an e.e. greater than zero?

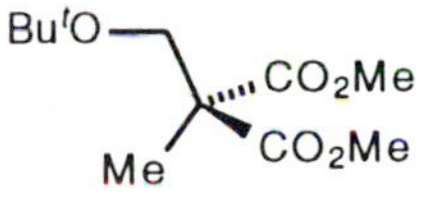

1.4

1.2 Classify the following reactions as either stereoselective or stereospecific:

(i) Reaction of epoxide **A1** with an achiral nucleophile.
(ii) Epoxidation of **A2** with *m*CPBA (a peroxyacid).
(iii) Epoxidation of **A3** with *m*CPBA.

*m*CPBA: meta-chloroperoxybenzoic acid.

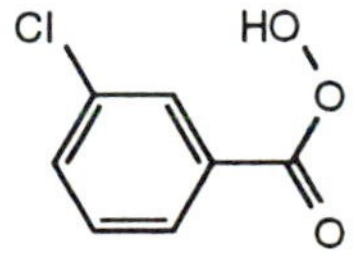

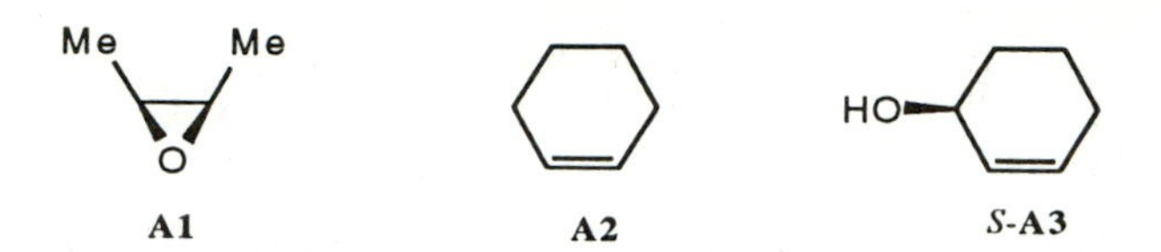

Suggestions for further reading

The principles of stereoisomerism and assignment of absolute configuration can be found in any modern basic organic text, e.g. J. McMurry, *Organic Chemistry*, 4th edn, Brooks/Cole, Pacific Grove, 1996, pp. 294–341.

A comprehensive treatment of stereochemistry, associated terms, and an introduction to stereoselective synthesis: E. L. Eliel, S. H. Wilen, and L. N. Mander, *Stereochemistry of Organic Compounds*, Wiley, New York, 1994.

2 Stereochemistry of reactions

2.1 Making and breaking bonds

General considerations

In Chapter 1 we considered some important aspects and consequences of the stereochemistry of organic compounds. This included the stereochemistry of reactants, products, catalysts, and even transition states. We viewed these, even transition states, as being 'static'. Perhaps it is more appropriate to say that we viewed these as not changing with time, at least on the time scale which we were representing them for our purposes. A reaction is completely different. The system must necessarily change with time, so what can we say about the stereochemistry of a reaction? Obviously we can consider the stereochemistry of the reactants, products, catalysts, transition states, and how these influence each other. These are important, and we will deal with them later. Here we are concerned with the reactions themselves.

It might seem on the face of it that a reaction cannot 'have' a stereochemistry, so first of all we must be clear what we mean. A reaction involves making and breaking of bonds, and this comes about by the interaction of one species with another, such as addition of a nucleophile to a carbonyl group. Bond formation occurs by the interaction of electrons and orbitals of both components. When two such components interact, in principle all the electrons and orbitals in each will interact to some extent, and at first sight it might be thought to be too complicated to analyse. Fortunately we can understand many important types of reactions using a very simple model, based on just one orbital from each component. Before discussing this model, we will consider the important types of orbitals involved. You might already be well acquainted with the following much simplified analysis, in which case you could proceed to Section 2.3. If it is all new to you, then you are strongly advised to consult a text which contains a much more detailed treatment (see further reading, especially Primer 26).

Bonding and antibonding molecular orbitals

The two most important types of orbitals which we need to analyse for reactions in organic chemistry are σ bonds, π bonds, and their respective antibonding orbitals. We will consider briefly C–C σ bonding and antibonding orbitals (C–C σ orbitals and C–C σ* orbitals).

These orbitals can be understood by using an imaginary process in which we combine two carbon radicals (both 'sp^3 hybridized'). In each radical, we imagine the unpaired electron to occupy an sp^3 orbital. The σ bond is formed by bringing these two radicals together (Fig. 2.1).

σ Bond: Formed by 'end -on' overlap of two orbitals.

π Bond: Formed by 'side-on' overlap of two orbitals.

Antibonding orbitals are usually denoted using the superscript '*'.

There are two possible ways in which these orbitals overlap. If they overlap with the same phase ('in phase combination') then a bonding orbital is formed, the C–C σ orbital. The alternative 'out of phase combination' has a nodal plane between the C atoms, and is the antibonding or C–C σ* orbital. Simple representations of these orbitals are shown in Fig. 2.1 (the bonding electrons have been omitted).

Nodal plane; A plane in an orbital in which the probability of finding electron density is zero. 'Node' is often used in place of 'nodal plane'. It can be viewed a result of a phase change in the orbital.

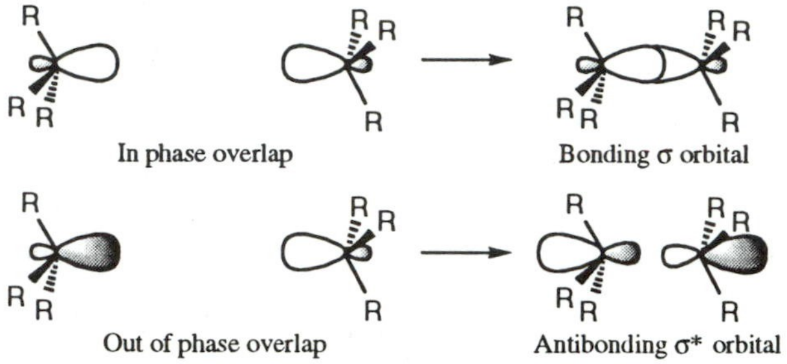

In phase overlap — Bonding σ orbital

Out of phase overlap — Antibonding σ* orbital

Fig. 2.1 Construction of bonding and antibonding σ orbitals

In the preceding discussion we considered an 'end-on' overlap of orbitals, which results in σ and σ* orbitals. A similar diagram can be constructed for the 'side-on' overlap of p orbitals. This is illustrated in Fig. 2.2, in which we imagine bringing two such p orbitals together to form two new molecular π orbitals, the bonding π and antibonding π* orbitals. This representation would correspond to a simple, symmetrical, alkene. Double bonds can, of course, form extended conjugated π systems when the double bonds are adjacent. The ideas which have been used so far can also be applied to such π systems.

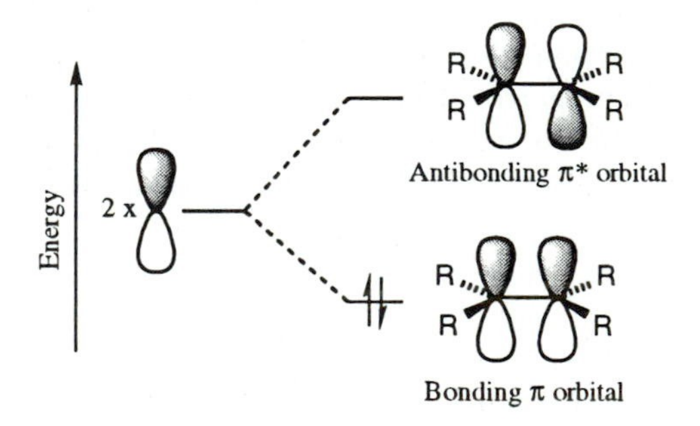

Fig. 2.2 Construction of π and π* molecular orbitals for a simple alkene

Dienes in which the double bonds are conjugated have a total of four p orbitals which interact, and this results in four π molecular orbitals, two bonding (π) and two antibonding (π*) orbitals (Fig. 2.3). Each of the p orbitals contained one electron, so we have four electrons to place in the π molecular orbitals. The lowest energy corresponds to pairs of electrons in the two bonding orbitals, and none in the antibonding orbitals.

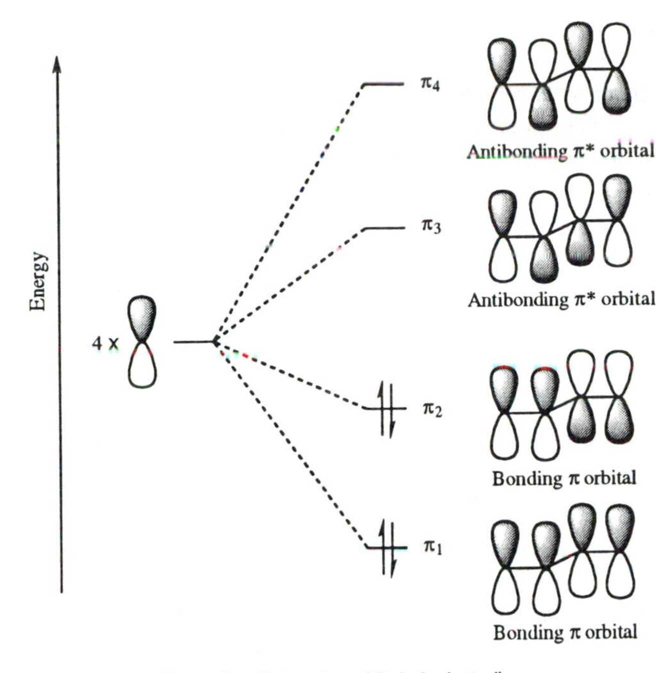

Fig. 2.3 π Molecular orbitals for butadiene

No more than two (paired) electrons can occupy any single molecular orbital.

By analogy, for a conjugated triene, in which six p orbitals interact, we might expect six π molecular orbitals, three bonding and three antibonding. This is the case, and the six electrons occupy the three bonding π molecular orbitals in pairs, and the antibonding orbitals are unfilled. The most important π orbitals from the point of view of reactivity are the highest bonding and the lowest antibonding orbitals. The π molecular orbitals of alkenes, including those of conjugated systems, are of particular importance in understanding many cycloaddition reactions (Chapter 6) and sigmatropic rearrangements (Chapter 9). These types of reactions are often highly stereoselective (or stereospecific) and are of great value in organic syntheses.

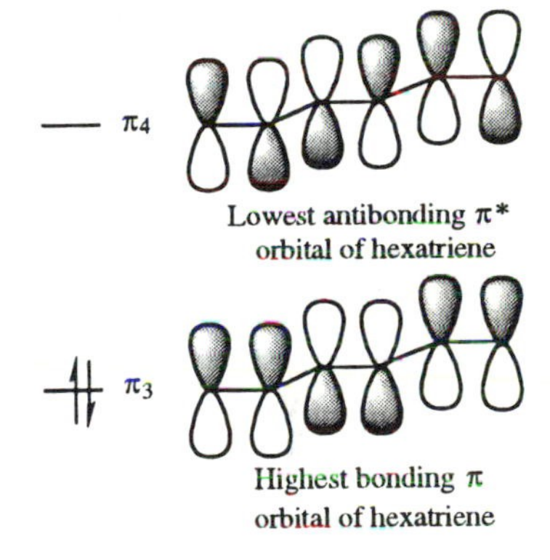

2.2 Orbital interactions and reactions

In section 2.1 we used a simple pictorial model to investigate bonding and antibonding orbitals of σ and π bonds. Reactions involve (usually) the interaction between two neutral molecular species, or a neutral molecular species and a cation or anion. We can use a similar pictorial model to understand the stereochemical consequences of the electronic aspects of these types of reactions, often referred to as stereoelectronic effects.

Stereoelectronic effects are dealt with in detail in Primer 36.

When two molecules approach each other, the molecular orbitals interact. We need to identify those interactions which might be crucial in determining the outcome of the reaction. Filled orbitals of one molecule interacting with filled orbitals in another leads to an increase in energy.

The interaction of two empty orbitals is usually irrelevant since they contain no electrons and so cannot affect the overall energy. The key orbitals will be the HOMO of one molecule (a filled orbital) and the LUMO of the other (an empty orbital).

HOMO - Highest occupied molecular orbital
LUMO - Lowest unoccupied molecular orbital

Let us consider a hypothetical reaction between **A** and **B**. What we need are orbital interactions between **A** and **B** which result in two new orbitals populated by only two electrons. This can be achieved if one filled orbital and one unfilled orbital interact (Fig. 2.4).

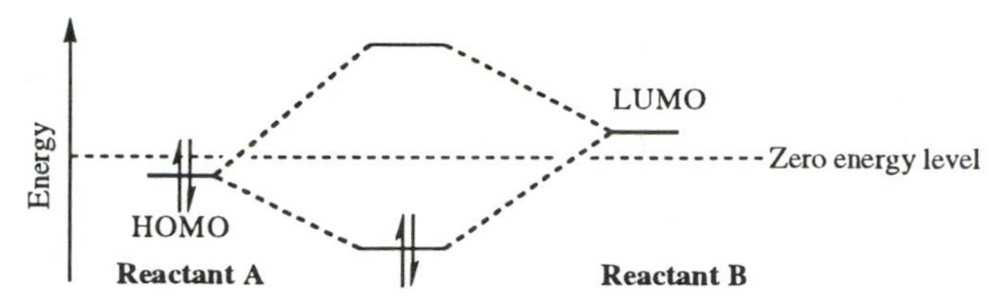

Fig. 2.4 Interaction of HOMO and LUMO of hypothetical reactants **A** and **B**

We now have a very simple method for identification of the orbital interactions which are likely to be important in understanding stereoelectronic effects in reactions. This process can be summarized thus. Identify the LUMO and HOMO of each reactant and establish which pair of HOMO/LUMO interactions result in greatest stabilization (i.e. which are closest in energy). This approach to the analysis of reactivity is often referred to as the frontier molecular orbital or FMO approach.

2.3 Stereoelectronics of some common reactions

S_N2 displacements

Nucleophilic substitutions which follow the S_N2 mechanism involve reaction of a nucleophilic species with a tetrahedral carbon atom carrying a leaving group. As we saw in Chapter 1 the reaction takes place strictly with inversion of the stereochemistry of the carbon atom which is undergoing the reaction. It is stereospecific, and it might be thought that we have now said all that can be said about the reaction. However, we can ask a number of further interesting questions. For example, which orbitals are involved, what are their shapes, how is maximum overlap achieved, and are there any stereoelectronic requirements?

Take a typical S_N2 reaction (Fig. 2.5), that of a secondary mesylate **2.1** with azide. The new bond is formed by donation of an electron pair from the nucleophile (N_3^-). In this case this is the filled orbital which we must consider. The nucleophile corresponds to reactant **A** in our earlier general analysis (Section 2.2). Mesylate **2.1** must correspond to reagent **B**, but which empty orbital is involved? It is the antibonding σ orbital (σ*) of the bond which is breaking. Having identified the HOMO and LUMO which are interacting, we can now construct a simple FMO picture corresponding to an early stage of the bond forming process (Fig. 2.5).

The best overlap will be achieved when the nucleophile and the two atoms comprising the σ bond undergoing cleavage are colinear. The bond which is forming and that which is breaking will be at an angle of 180°. This is illustrated in Fig. 2.6, and can be generalized to include all S_N2 reactions.

N_3^- ---- R, Me, H, 180° —OMs → [N_3---- R, Me H, 180° ----OMs]$^-$ → N_3— R, H, Me, 180° ----→ $^-$OMs

Nuc^- ---- a, b, c, 180° —L.Gp. → [Nuc---- a, b c, 180° ----L.Gp.]$^-$ → Nuc— a, c, b, 180° ----→ L.Gp.$^-$

Nuc^- = nucleophile L.Gp. = Leaving group

Fig. 2.5 Typical S_N2 reaction

Stereoelectronic considerations and requirements, far from being minor and somewhat abstruse aspects of organic reactions, can play a decisive role in determining reaction pathways and mechanisms. Bearing this in mind, we will consider reactions of unsaturated systems, C=C and C=O.

N_3^- ---- R, Me, H, 180° —OMs → [N_3---- R, Me H, 180° ----OMs]$^-$ → N_3— R, H, Me, 180° ----→ $^-$OMs

Nuc^- ---- a, b, c, 180° —L.Gp. → [Nuc---- R, Me H, 180° ----L.Gp.]$^-$ → Nuc— a, c, b, 180° ----→ L.Gp.$^-$

Nuc^- = nucleophile L.Gp. = Leaving group

Fig. 2.6 Stereoelectronic requirements for an S_N2 reaction

Additions to simple alkenes

Simple alkenes do not react readily with most nucleophiles, they are in fact slightly nucleophilic themselves, and as such their (ionic) reactions are dominated by electrophilic additions. The orbitals which we need to consider are again easy to determine. The alkene acts as the nucleophile so we need the HOMO of the alkene, and consequently the LUMO of the electrophile. Maximum overlap will occur when the electrophile approaches the alkene as shown in Fig. 2.7.

From this simple analysis, we expect that when an electrophilic species attacks a double bond, both new bonds will be formed from the same face of the alkene. Overall the reaction should be a *cis* addition. A typical example

The electrophile LUMO can interact strongly with both p orbitals of the alkene HOMO.

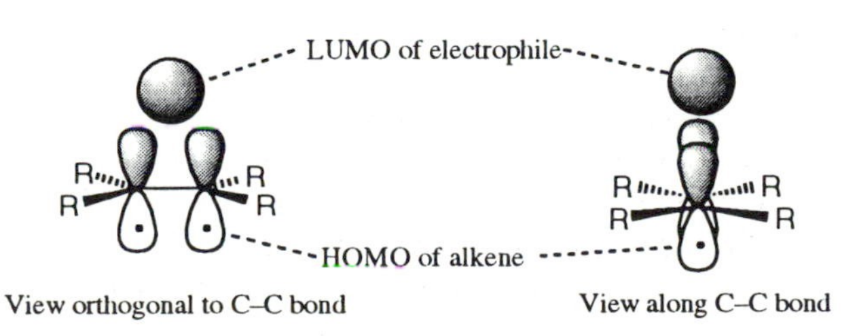

Fig. 2.7 Stereoelectronic requirements for electrophilic addition to and alkene

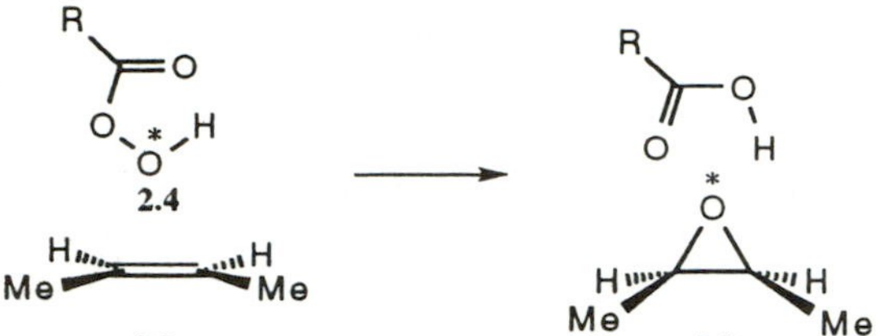

Fig. 2.8 Epoxidation of an alkene

of such a reaction is the reaction of an alkene with a peroxyacid to produce an epoxide, such as that shown in Fig. 2.8. In this reaction, the epoxidation of *cis*-2-butene **2.2**, the stereochemistry of the alkene is reflected in the product epoxide **2.3**. The electrophile is effectively the oxygen of the peroxyacid (**2.4**) which is labelled O*, as this is the one which is transferred.

Other electrophilic reactions are slightly more complicated as they involve the addition of two atoms to the alkene. A typical example is the addition of bromine to an alkene. In this case the reaction is a two step process. The bromine molecule is the electrophile, and reacts initially to produce a cyclic bromonium ion, analogous to the epoxide produced by epoxidation of **2.2** (Fig. 2.8). This bromonium ion then reacts with the bromide ion which resulted from the initial electrophilic attack. This second reaction can be viewed as an S_N2 displacement, with the Br^+ of the bromonium ion acting as the leaving group. Following the stereoelectronic requirement for S_N2 displacements, this should take place so that the forming and breaking bonds are colinear, we can predict the overall stereochemical outcome of the addition of bromine to an alkene as for *cis*-2-butene **2.2** in Fig. 2.9.

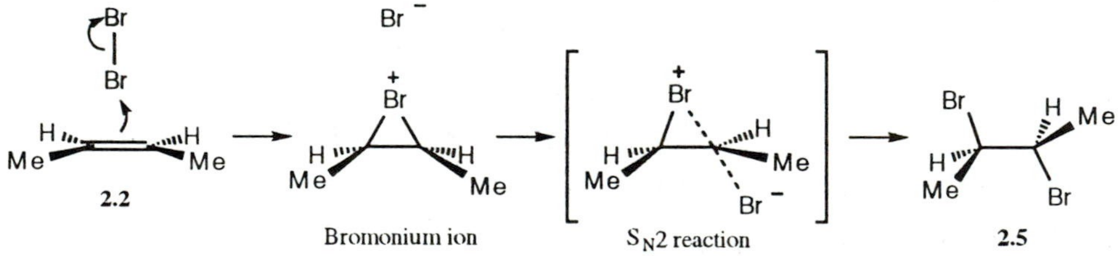

Fig. 2.9 Stereochemistry of addition of bromine to an alkene

Inspection of the product **2.5** reveals that the stereochemistry of the alkene is 'preserved' in the product and the bromine has added in a *trans* manner.

Some electrophilic additions (of two atoms) to alkenes take place with overall *cis* stereochemistry. In the case of hydroboration, one of the most widely used examples of this type, the first step in the reaction is formation of a π complex, which would be analogous to the bromonium ion in Fig. 2.9. This leads to a four-centre transition state as illustrated in Fig. 2.10, and ultimately to an overall *cis* addition.

Additions to the carbonyl group

The carbon atom of a carbonyl group is electrophilic, and addition of a nucleophile is a particularly important reaction is organic synthesis. By

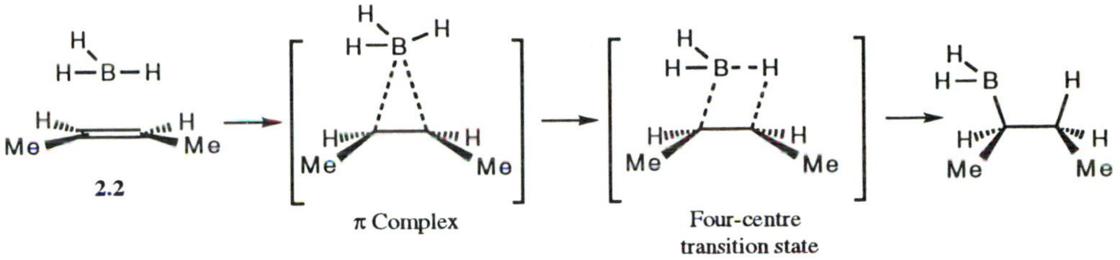

2.2

π Complex

Four-centre transition state

Fig. 2.10 Stereochemistry of hydroboration of an alkene

analogy with our simple FMO approach used in the foregoing sections we need to examine the HOMO of the nucleophile and the LUMO of the carbonyl group. An orbital picture for addition of a nucleophile to a carbonyl is shown in Fig. 2.11. It can be seen that the best angle of attack of the nucleophile is not ninety degrees, as might be imagined. Attack at ninety degrees would indeed result in maximum overlap between the HOMO of the nucleophile and the p orbital centred on the carbon atom of the carbonyl group. However there will be significant antibonding (out-of-phase) overlap with the other p orbital which makes up the LUMO (π*) of the carbonyl group. The best compromise between these two opposing effects is achieved if the nucleophile attacks at an angle of approximately 109.5°. This is maintained between the forming bond and the breaking bond throughout the reaction.

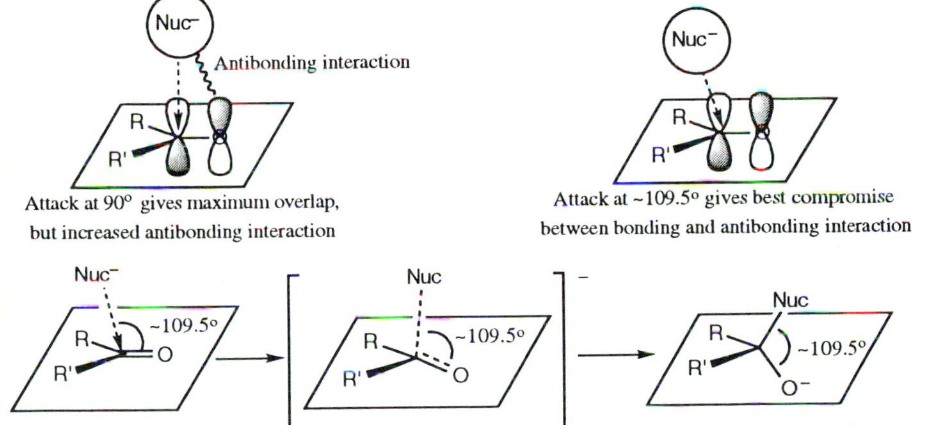

Fig. 2.11 Stereoelectronic requirements for nucleophilic addition to a carbonyl group

The argument used to account for the approach direction of a nucleophile to a carbonyl group can be extended to those reactions which involve nucleophilic addition to C=C bonds. This reaction takes place when the C=C bond is electron deficient, usually when conjugated to a carbonyl group or other strongly electron withdrawing group. This will be discussed later (Chapter 6). Almost all the reactions dealt with so far have been ionic. Concerted processes, in which bonds are made and broken more or less simultaneously, are also of great value in stereoselective organic synthesis, and we will now see how our simple FMO picture can account for some of the important aspects of Diels–Alder cycloadditions and Claisen/Cope rearrangements.

Diels–Alder cycloadditions

This reaction is one of the most powerful reactions in organic synthesis. It involves reaction of a diene and an alkene (known as the dienophile) to form a six membered ring. In the reaction four sp^2 centres, two each from the diene and dienophile, become tetrahedral (Fig. 2.12).

1 2 3 4 + Dienophile → 1 2 3 4 ● = sp^3 Centres formed in the reaction

Fig. 2.12 Schematic Diels–Alder cycloaddition

This means that there is the potential for the creation of four new stereocentres in the product. This reaction will be discussed in more detail in Chapter 6, and here we will be concerned with illustrating the important aspects of the reaction and the orbitals involved.

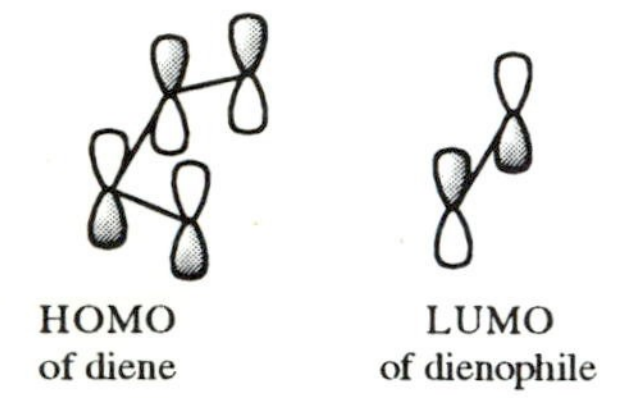

The reaction clearly involves only π electrons, and is known to be a thermal, concerted process. To apply our simple FMO analysis, we need to decide which HOMO-LUMO pair to invoke. Normal Diels–Alder cycloadditions involve electron deficient (and therefore electrophilic) dienophiles, so we will take the LUMO of the dienophile. We have encountered the HOMO of the diene earlier (Fig. 2.3). The HOMO and LUMO which are involved are illustrated in Fig. 2.13 (the diene is drawn in the conformation which it must adopt for reaction).

Best overlap will be achieved if the interacting p orbitals (those which will ultimately lead to the two new σ bonds) overlap ‘end on’, leading to a transition state analogous to **2.6** (Fig. 2.13).

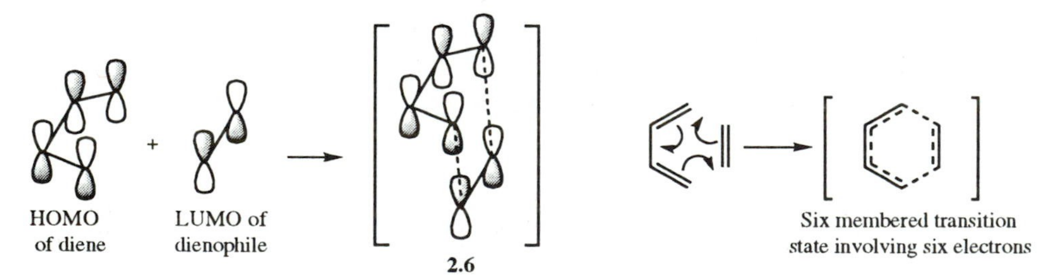

Fig. 2.13 Transition state for the Diels–Alder reaction

From the point of view of the dienophile, this is a *cis* addition and we should expect that stereochemistry present in this component should be preserved in the product. This is indeed the case, as can be seen from the reaction of butadiene **2.7** and dienophile **2.8** (Fig. 2.14). The ester groups are *trans* in the dienophile and in the product. Further discussion of the Diels–Alder cycloaddition and related reactions can be found in Chapter 6.

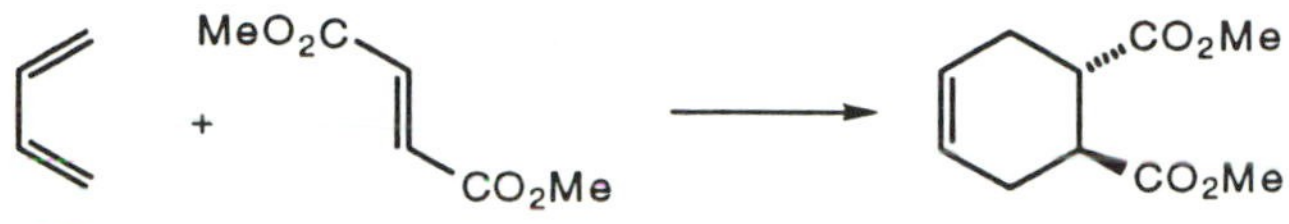

Fig. 2.14 A typical Diels–Alder cycloaddition

Claisen/Cope rearrangements

These two reactions are closely related, and also related to the Diels–Alder cycloaddition. Like the Diels–Alder reaction, these rearrangements are essentially concerted, thermal reactions which involve six electrons and a six-membered transition state (Fig. 2.15).

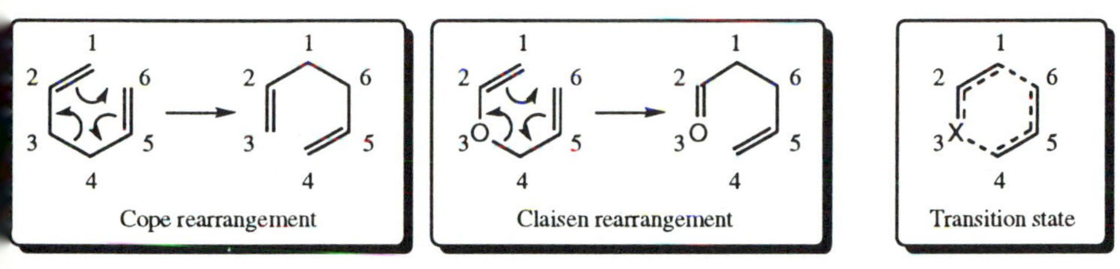

Fig. 2.15 The Claisen and Cope rearrangements

To apply our simple FMO picture to this problem, we need to adopt a somewhat different approach, as there is only one reactant. Inspection of the transition state (Fig. 2.15) shows that atoms 1, 2, and 3 remain attached to each other, as do atoms 4, 5, and 6. For our simple FMO approach, we can imagine these units each to be 'reactants'. Furthermore we will imagine that one unit (atoms 1, 2, and 3) to be an anion, and the other to be a cation. This is *not* the mechanism, just an artificial device so that we can apply our simple model to understand the stereochemistry of the reaction. The reaction itself is concerted and not ionic.

We are taking the unit containing atoms 1, 2, and 3 to be an anion, so we would expect to use its HOMO, and therefore to use the LUMO of the other unit (atoms 4, 5, and 6). Both these units are allyl systems, and the π molecular orbitals for such systems are illustrated in Fig. 2.16.

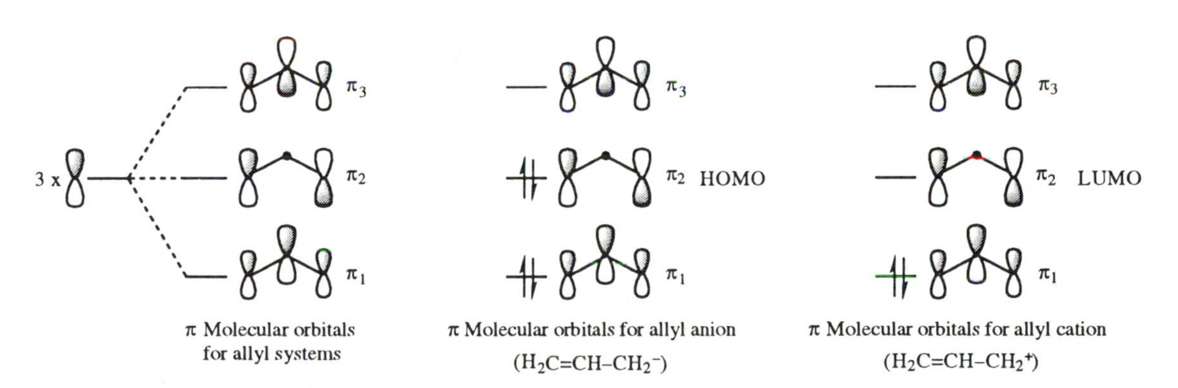

Fig. 2.16 π Molecular orbitals for allyl systems

The orbitals which we need are the HOMO of the allyl anion and the LUMO of the allyl cation. If we now use them to 'build' an orbital picture corresponding to the transition state (Fig. 2.15), then we arrive at a transition state similar to **2.9** (Fig. 2.17). A representation of this transition state with only the six-membered ring and the forming and breaking bonds (dashed lines) **2.10** indicates that the stereochemistry of this transition state in similar to that of a cyclohexane molecule in the chair conformation. The conventional representation of this is **2.11**.

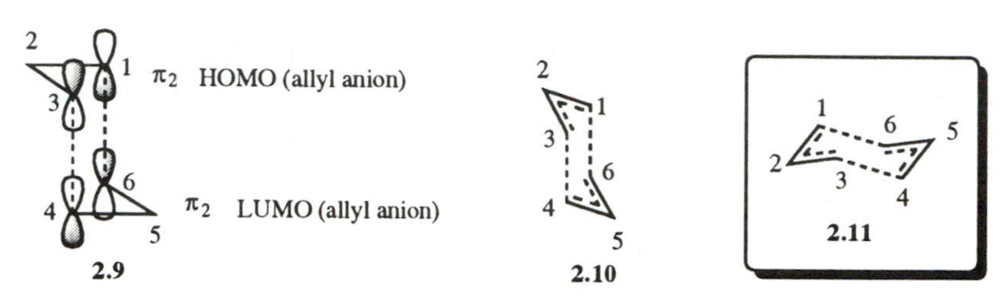

Fig. 2.17 Transition state for Claisen and Cope rearrangements

This is an important type of transition state (which is why we have spent some time on it!) which we will encounter when discussing additions to carbonyl compounds and aldol reactions (Chapters 3 and 5 respectively), as well as Claisen rearrangements (Chapter 9).

Detailed discussion of the stereochemical implications of this type of transition state will be delayed until the appropriate chapter, but we will consider here some aspects which are common to the various reaction types which involve such 'chair' transition states. The atoms which started out as sp^3 centres, and those which are becoming sp^3 centres are treated as if they are tetrahedral. In our general structure this would correspond to atoms 1,3,4, and 6. The transition state is imagined to be like cyclohexane which means that each of these atoms could carry substituents which might be axial or equatorial. This analogy between cyclohexane **2.12** and the transition state **2.13** for Cope rearrangement of 1,5-hexadiene **2.14** is shown in Fig. 2.18.

Further discussion will be deferred to the appropriate sections, but an

2.14

2.13

Six membered ring chair transition state (H* = equatorial)

2.12

Chair conformation of cyclohexane (H* = equatorial)

Fig. 2.18 Chair representation of transition state for Claisen and Cope rearrangements

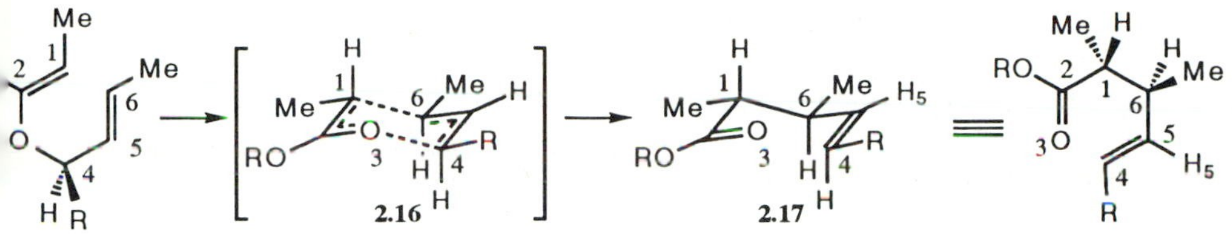

Fig. 2.19 A typical Claisen rearrangement

example of the power of this type of reaction is presented in Fig. 2.19. This involves the Claisen rearrangement of compound **2.15**, as a single enantiomer. The group R is very bulky, and the lowest energy for the transition state will be when this group at C-4 is equatorial. The groups at C-1 and C-6 can then also be equatorial in conformation **2.16**. The product would then be expected to correspond to **2.17**, which is found to be the case.

Eliminations to form C=C bonds

All the reactions which we have covered so far have involved either the conversion of one stereocentre into another (e.g. S_N2 reactions), or the creation of new stereocentres in the product. Although they will not be covered in much detail in this text, the formation of alkenes with a particular geometry often involves the conversion of tetrahedral centres into trigonal (sp^2) carbon atoms. This process is typified by the elimination of H–X from adjacent sp^3 centres in the presence of base (X = good leaving group).

There are two extreme mechanisms which are observed for this type of elimination. One involves two discrete steps (E1cb). The first step, which is not rate determining, is the removal of the proton to give an intermediate carbanion (still sp^3). This is stable enough to allow conformational equilibria to be established. In the case of carbanion formed from **2.18** on treatment with base, there are three staggered conformations **2.19**, **2.20**, and **2.21**.

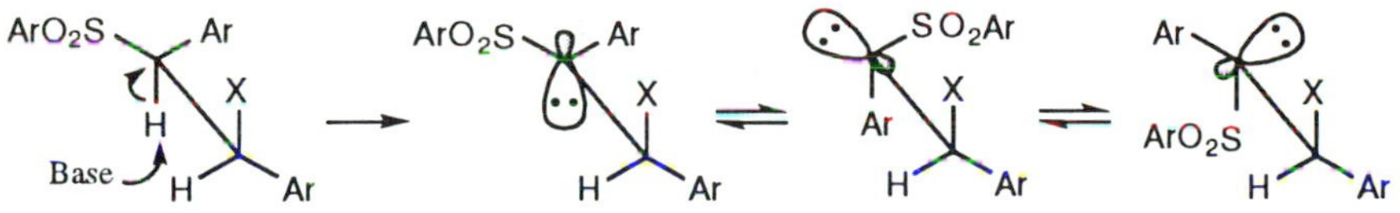

Fig. 2.20 Staggered conformations for an E1cb elimination

If we now consider the elimination to be similar to an 'internal' S_N2 displacement, the sp^3 'anion' becomes the nucleophile and the group X the leaving group. The orbitals we need to examine then become the HOMO of the 'anion' (its σ orbital) and the LUMO of the C–X bond (its σ* orbital). Maximum overlap will be achieved when these two orbitals are *syn* and coplanar (for obvious geometric reasons they can never become colinear as in a 'proper' S_N2 displacement). This places the C–X bond antiperiplanar to the 'anion' (Fig. 2.21).

For the anion from **2.18** (Fig. 2.20) only one of the three staggered conformations satisfies this stereoelectronic requirement, conformation **2.19**.

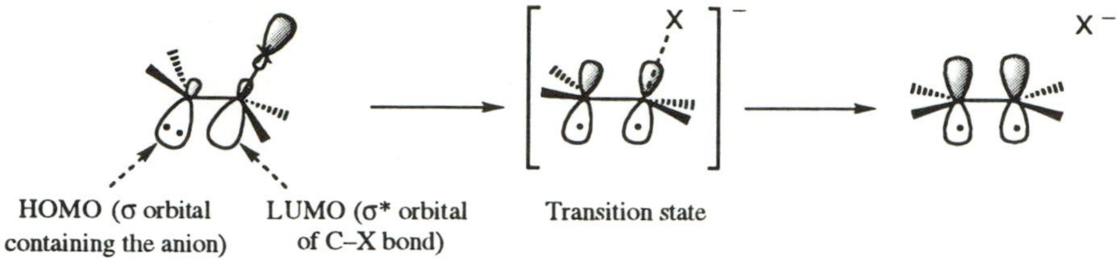

HOMO (σ orbital containing the anion) LUMO (σ* orbital of C–X bond) Transition state

Fig. 2.21 Stereoelectronic requirements for an E1cb elimination

This leads to the expectation that the geometric isomer of the alkene produced should be **2.22**, which is in fact the experimental observation (Fig. 2.22).

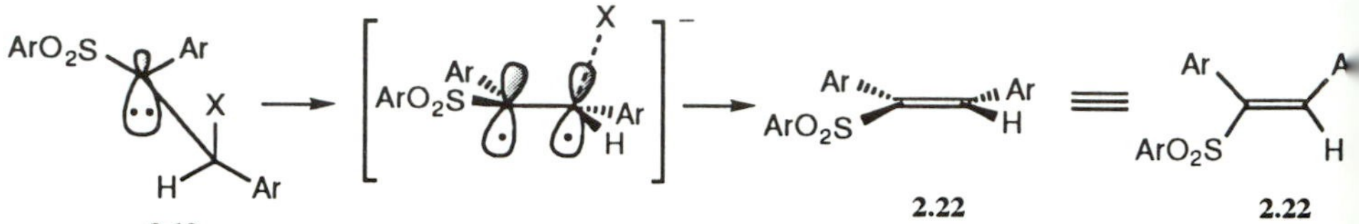

Fig. 2.22 Stereochemistry of alkene formed in an E1cb elimination

The other extreme for this type of elimination involves a concerted removal of the proton and loss of the leaving group (E2). The C–H and C–X bonds are broken more or less simultaneously.

By analogy with the preceding discussion, the HOMO corresponds to the σ orbital of the C–H bond, and the LUMO to the σ* orbital of the C–X bond. The preferred conformation again has the C–H and C–X bonds antiperiplanar resulting in a *trans* elimination Fig. 2.23). As can be seen from Fig. 2.23, if we label the substituents on the two carbon atoms a, b, c, and d, then the alkene produced must have the geometry shown.

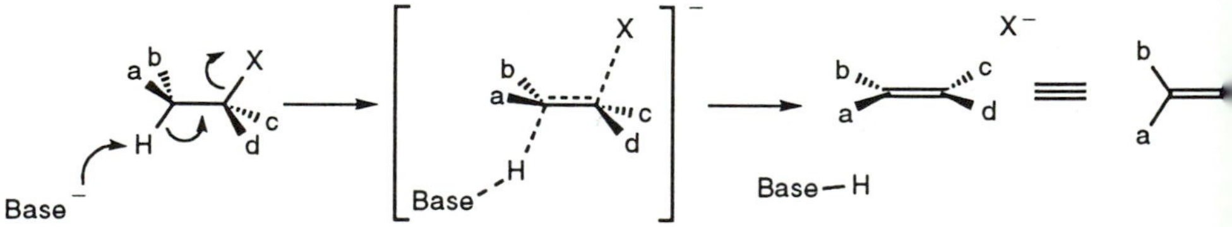

Fig. 2.23 Stereoelectronic requirements for an E2 elimination

In this chapter, we have seen how we can use simple ideas relating to the orbitals which are involved in a wide range of reactions to predict or account for the stereochemical outcome of the reactions. This is a very important idea, which is why we have spent a considerable amount of time on the topic. In the following chapters we will be using these ideas, where appropriate, as an aid to understanding the stereochemistry of some of the most important types of reactions which are encountered in organic chemistry.

2.4 Problems

2.1 Predict the stereochemical outcome of the following reactions:

Me Me + Br_2 **B1**

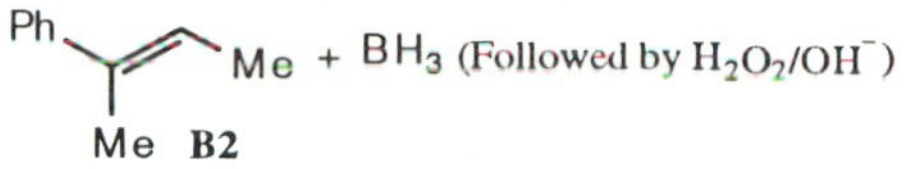

Ph Me Me + BH_3 (Followed by H_2O_2/OH^-) **B2**

2.2 Account for the observation that the diastereoisomeric 2-bromo-4-phenylcyclohexanols shown below react differently on treatment with base.

Hint: Make molecular models and examine which reactions are feasible.

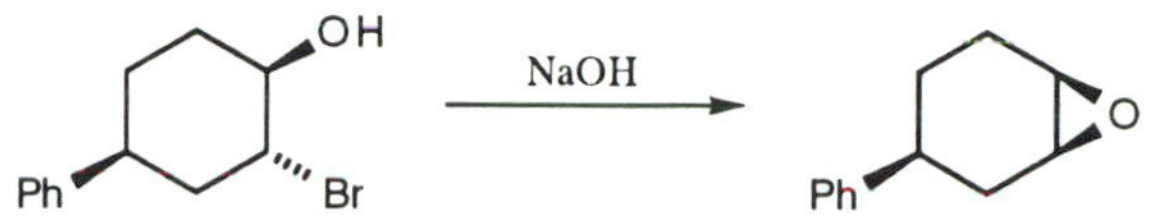

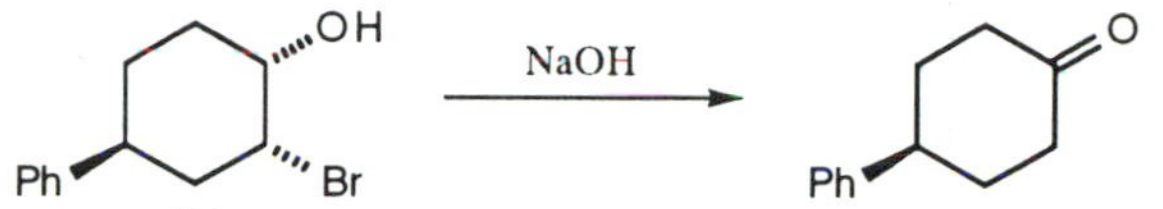

Ph OH Br **B3** → NaOH → Ph O **B5**

Ph OH Br **B4** → NaOH → Ph O **B6**

Suggestions for further reading

W. G. Richards and P. R. Scott, *Energy Levels in Atoms and Molecules*. Oxford Chemistry Primer no. 26, Oxford University Press, Oxford.

Orbitals and reactions: I. Fleming, *Frontier Orbitals and Organic Chemical Reactions*, Wiley, New York, 1977; T. A. Albright, J. K. Burdett, and M. H. Whangbo, *Orbital Interactions in Chemistry*, Wiley, New York, 1985. A. J. Kirby, *Stereoelectronic Effects*, Oxford Chemistry Primer no. 26, Oxford University Press, Oxford, 1996; P. Deslongchamps, Stereoelectronic Effects in Organic Chemistry, Pergamon Press, Oxford, 1983; A. J. Kirby, *The Anomeric Effect and Related Stereoelectronic Effects at Oxygen*, Springer–Verlag, Berlin, Heidleberg, and New York, 1983.

3 Additions to carbonyl compounds

3.1 The reaction

The general reaction with which we will be concerned in this chapter is illustrated in Fig. 3.1, and represents one of the most important reactions in organic synthesis (the related aldol reaction, addition of an enolate to an aldehyde or ketone, is dealt with in Chapter 5).

Priority order O>R^1>R^2 —Nuc→ [Nucleophile attacking from the *re* face] (109.5°) - - - → (*S*)-Enantiomer formed (assuming priority order O>R^1>R^2>Nuc)

Fig. 3.1 Attack of a nucleophile on a carbonyl group

The fundamentals of additions to carbonyl compounds have been covered in the previous two chapters. A summary of these is presented here. A carbonyl group is prochiral if $R^1 \neq R^2$, and the following apply:–

For the definition of *re* and *si*, see Chapter 1.

(a) The two faces can be assigned as either *re* or *si*.
(b) The *re* and *si* faces are enantiotopic if the molecule is not chiral.
(c) The *re* and *si* faces are diastereotopic if the molecule is chiral.
(d) The nucleophile approaches the carbonyl group so that an angle of approximately 109.5° is maintained between the nucleophile and the breaking C=O bond.

Nucleophilic additions to carbonyl groups can be divided into those which give enantiomeric products, and those whose products are diastereoisomers. In the latter case, the carbonyl compound is often chiral, and we will consider this class of additions first.

3.2 Additions to chiral carbonyl compounds

For simple, acyclic, carbonyl compounds such as **3.1** containing one stereocentre (Fig. 3.2), we would expect the two diastereoisomeric products **3.2** and **3.3** to be formed in different amounts. Given that there is 'free' rotation about the C–C bond which connects the stereocentre with the carbonyl group it might be expected that the diastereoselectivity would be low. This is not necessarily the case, and with the appropriate carbonyl compound and nucleophile the diastereoselectivity can be remarkably high.

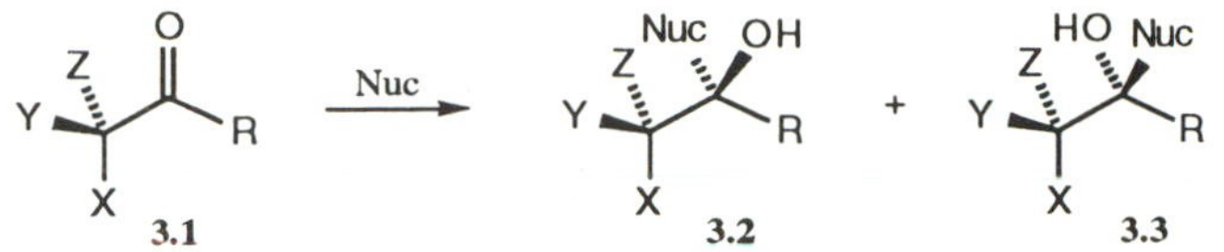

Fig. 3.2 Addition to a chiral ketone or aldehyde

If one of the groups X, Y, or Z, is capable of being involved in chelation of a metal counter-ion, along with the oxygen of the carbonyl group, then the stereoselectivity and the model used to account for it changes. This type of reaction will be considered following our discussion of 'simple' chiral carbonyl compounds.

Nucleophilic addition to simple carbonyl compounds

There have been several models advanced to account for the trends observed in the diastereoselectivity of nucleophilic additions to ketones and aldehydes such as **3.1**. Such models need to take into account the various factors which might influence such additions. These include conformational effects, steric interactions, electronic factors, and stereoelectronic effects.

We will consider only the model commonly referred to as the 'Felkin–Ahn' model. An important tenet of this model is that nucleophilic attack takes place antiperiplanar to a neighbouring σ bond (Fig. 3.3).

This conformation is thought to be favoured because of overlap of the neighbouring antibonding (σ*) orbital of the σ bond with the π* orbital of the carbonyl group. In the absence of the nucleophile both these orbitals are unfilled, and therefore their interaction will have no effect on the energy. However, as the nucleophile attacks, electron density is transferred from the electron rich nucleophile to the π* orbital of the carbonyl group. When there is a neighbouring σ* orbital suitably aligned then the electron density which is building in the π* orbital of the carbonyl group can be 'delocalized', resulting in an overall lowering in energy (Fig. 3.3).

Nucleophilic attack is favoured when a neighbouring σ bond is antiperiplanar. In our typical example **3.1** there are three such bonds, C–X, C–Y, and C–Z, so which do we choose?

If we place the largest group (X, Y, or Z) antiperiplanar, then we will minimize steric repulsions. However, if there is an electronegative group amongst X, Y, and Z, which is incapable of chelation (see below) then it is placed antiperiplanar to the nucleophile. This helps to reduce electrostatic repulsion. Moreover, the corresponding σ* orbital will be lower in energy and therefore will interact most strongly with the 'filling' π* orbital. On the

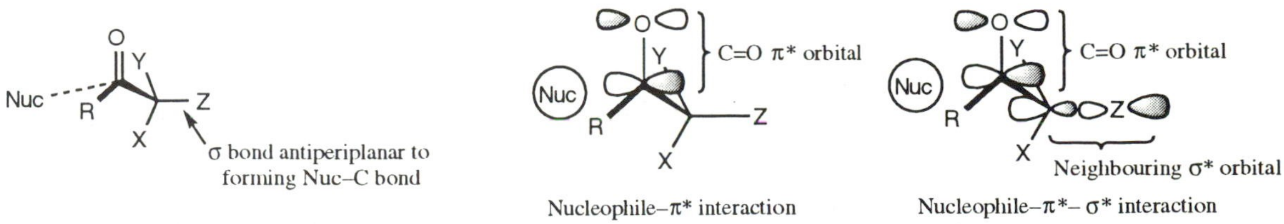

Fig. 3.3 Antiperiplanarity of the new bond and an adjacent σ bond in nucleophilic attack on a carbonyl group.

whole, we will be dealing with groups X, Y, and Z which are of similar electronegativity, and so usually we will be placing the largest substituent antiperiplanar to the incoming nucleophile.

It will be useful therefore to classify the three substituents X, Y, and Z, as small (S), medium (M), and, large (L). If we do this, then we narrow the choice of reacting conformations to two, **3.4** and **3.5** (Fig. 3.4). Nucleophilic attack on **3.4** is favoured over attack on **3.5**. In the latter, the nucleophile must pass close to the medium-sized group (M), whereas for **3.4** the nucleophile interacts with the smallest group.

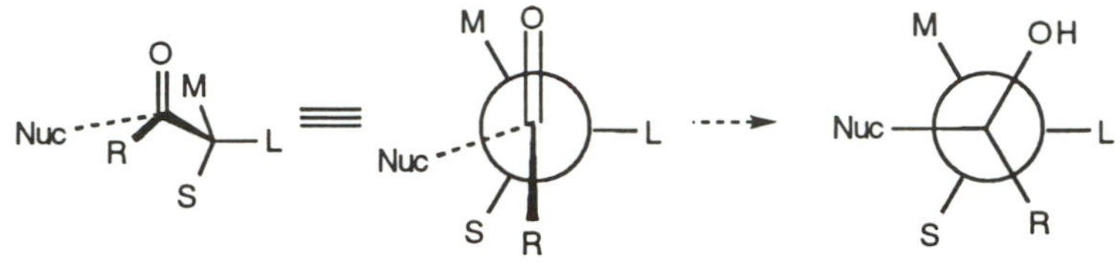

3.4 **3.6**

Nucleophile attacks C=O at ~109.5° so Nuc passes close to smallest group (S)

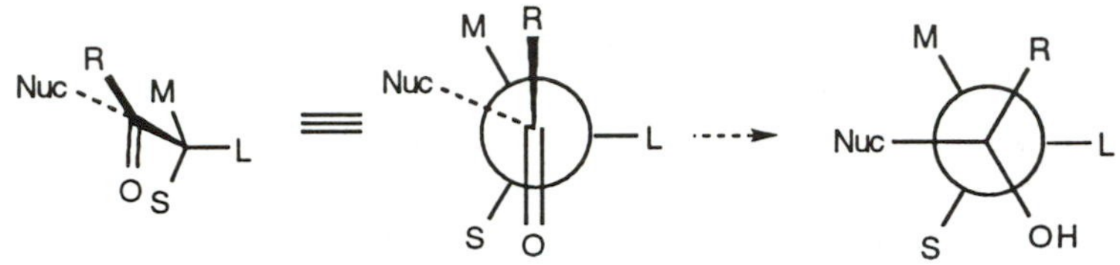

3.5 **3.7**

Disfavoured as Nuc attacks C=O at ~109.5° so Nuc passes close medium-sized group (M)

Fig. 3.4 'Felkin–Ahn' model for diastereoselective addition to a C=O group

Reactions on **3.4** and **3.5** take place on opposite faces of the carbonyl group, and so result in diastereoisomeric products **3.6** and **3.7**. If we assume that reaction takes place only on **3.4** and **3.5**, then the difference in energy between these will be reflected in the d.e. of the product (this also assumes that they are equally reactive).

Me(metal) = MeMgI, d.e. 33%
Me(metal) = $MeTi(OPh)_3$, d.e. 86%

d.e. 75%

Fig. 3.5 Simple examples of diastereoselective addition to ketones and aldehydes

In general, aldehydes react with lower stereoselectivity than ketones with the same reagent, and increasing the steric bulk of the reagent increases stereoselectivity. Choice of organometallic reagent can also have a dramatic effect, with titanium reagents offering excellent levels of stereoselectivity in some cases (Fig. 3.5).

If the carbonyl group is part of a ring, as in cyclohexanones, then this will clearly have a serious effect on the conformations which are accessible for reaction. The stereochemistry of addition to chiral cyclohexanones is rather complicated, and a clear, simple model has yet to emerge.

For a conformationally locked cyclohexanone, the two possible directions for nucleophilic attack can be labelled as 'axial' and 'equatorial' (Fig. 3.6). Relatively small nucleophiles (such as complex hydrides and alkyne anions) generally prefer 'axial' addition, but this preference can be overturned by the presence of axial groups at positions 3 and 5.

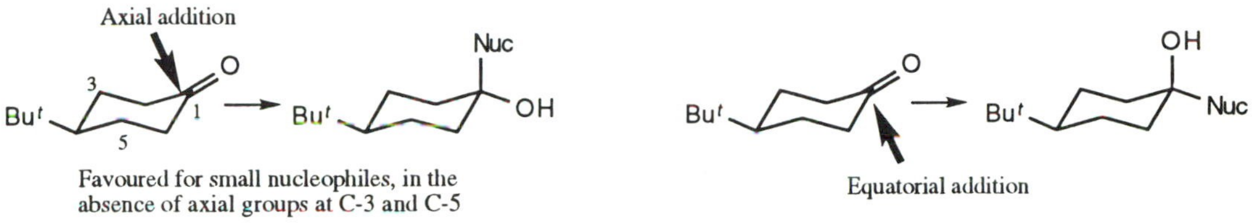

Fig. 3.6 Nucleophilic addition to cyclohexanones

Nucleophilic addition to carbonyl compounds capable of chelation

Nucleophilic reagents usually come associated with a metal counter-ion. If the carbonyl compound undergoing nucleophilic addition is capable of chelating this metal ion, then very high diastereoselectivity is possible. The chelation arises because one of the groups adjacent to the carbonyl group possesses a lone pair, usually either an oxygen or nitrogen substituent.

The carbonyl group also possesses lone pairs, so the carbonyl compound can act as a bidentate ligand for the metal ion. This has the effect of 'locking' the conformation, so maximizing the steric effects of the other two substituents adjacent to the carbonyl group (small, S, and large, L, Fig. 3.7).

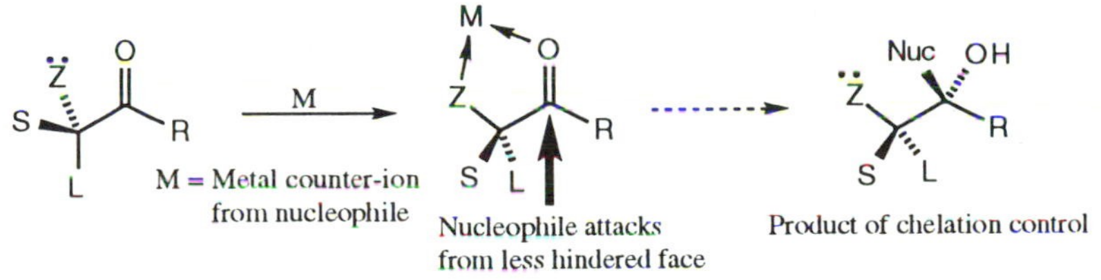

Fig. 3.7 'Chelation control' in the addition of nucleophiles to C=O groups

This type of reaction is often referred to as being under 'chelation control', and the level of stereoselectivity can be very high. For the reactions shown in Fig. 3.8, the d.e. of the product is 96 per cent or greater.

Fig. 3.8 Examples of 'chelation control' in additions to aldehydes and ketones

3.3 Additions to prochiral carbonyl compounds

If the carbonyl compound itself is prochiral, in most of the reactions of interest to us, another component of the reaction must be chiral. The most obvious candidate is the nucleophile. A simple example of this approach is the use of anions derived from chiral sulfoxides as nucleophiles.

Sulfoxides with two different groups attached to the sulfur atom are pyramidal, as there is also a lone pair on sulfur. They are configurationally stable up to ~200°C.

Chiral sulfoxides are configurationally stable, and can be prepared in essentially enantiomerically pure form. The sulfoxide group stabilizes an adjacent negative charge, so they can be deprotonated relatively easily with amide bases such as lithium di*iso*propylamide (LDA). The resulting chiral anions undergo stereoselective addition to prochiral carbonyl compounds, although the d.e. is often relatively low (Fig. 3.9).

LDA; $PhCOCF_3$

d.e. ~50%

Fig. 3.9 Addition of a chiral nucleophile to a prochiral ketone

An alternative approach to the use of chiral nucleophiles is to react a readily available achiral nucleophile with a chiral reagent, before carrying out the addition to the carbonyl compound. In effect, a 'standard' nucleophile, such as a Grignard reagent or an alkyl lithium, is converted in the reaction flask to a new chiral nucleophile. After reaction, the chiral reagent (or its equivalent) is removed either in the work up of the reaction or by standard purification techniques (see Chapter 1, section 1.5 for a discussion of the principles of using chiral reagents).

This approach is exemplified by the use of the titanium complex **3.8** (Fig. 3.10). Allylmagnesium chloride reacts with **3.8** to give the new chiral reagent **3.9**, and this reacts with benzaldehyde (and other aldehydes) with high enantioselectivity.

$CH_2{=}CHCH_2MgCl$ PhCHO

3.8 **3.9** e.e. 95%

Fig. 3.10 Addition of a chiral allyltitanium reagent to benzaldehyde

In the above example, the nucleophile is an allyl group. This type of nucleophile often reacts through a cyclic transition state, analogous to those of Claisen and Cope rearrangements (Chapter 2). Most of these reactions use allylic groups attached to boron, and a number of chiral boron reagents have been developed to control the stereochemistry of this type of reaction. Some important aspects of this type of reaction are illustrated in Fig. 3.11, along

Attack on *si* face of RCHO

Attacks on *si* face of RCHO

Attacks on *si* face of RCHO

Attacks on *re* face of RCHO

Ar = p-$CH_3C_6H_4$

Fig. 3.11 Chiral boron reagents for enantioselective addition to aldehydes

with some examples of allylic boron reagents which react with aldehydes with high enantioselectivity.

A feature of any chiral reagent, is that (at least) a mole of reagent is needed for each mole of carbonyl compound which reacts. As discussed in Chapter 1 (section 1.5) in principle using an asymmetric catalyst is much more efficient than either a chiral reagent or auxiliary. Such catalytic methods can be very effective in nucleophilic additions to carbonyl compounds. A relatively well studied method involves the use of an organozinc nucleophile and a chiral, enantiomerically pure 1,2-aminoalcohol as catalyst. The example used in Chapter 1 (section 1.5) and illustrated below is particularly effective. The catalyst, known as (+)- or (–)-DAIB **3.10**, is prepared from camphor, both enantiomers of which are readily available.

The acronym DAIB is derived from the compound's name, (+)- or (–)-3-*exo*-(dimethylamino)isoborneol.

(+)-DAIB (+)-**3.10**

(–)-DAIB (–)-**3.10**

The reaction is particularly effective for aldehydes. Nucleophilic addition of organozinc reagents to aldehydes is usually very slow, but the catalyst speeds up the addition.

Many types of aldehyde will react with organozinc reagents with catalysis by DAIB, and a selection of such reactions is presented in Fig. 3.12. There are other types of catalyst which can be used for the enantioselective addition of achiral nucleophiles to achiral carbonyl compounds. The most successful types contain 1,2-aminoalcohol units analogous to that in DAIB.

Rate enhancement is important if high levels of enantioselectivity are to be achieved. The rate of the 'uncatalysed' nucleophilic addition must be relatively slow

Me_2Zn, (–)-DAIB (2%) — e.e. 91%

$(C_5H_{11})_2Zn$, (–)-DAIB (2%) — e.e. > 95%

$(C_5H_{11})_2Zn$, (–)-DAIB (2%) — e.e. 96%

Et_2Zn, (–)-DAIB (2%) — e.e. 90%

Fig. 3.12 Enantioselective catalysis of nucleophilic addition to aldehydes

3.4 Problems

3.1 Which product should predominate in the reaction of **C1** with methyllithium?

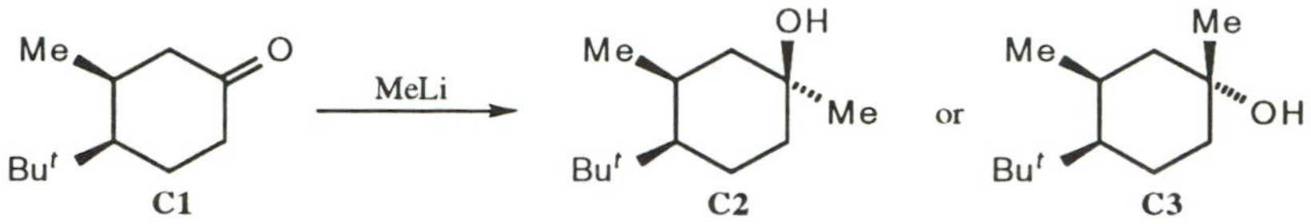

3.2 The reaction of dimethylzinc with **C4** is catalysed by (+)-DAIB ((+)-**3.10**). Which enantiomer of the product will be formed in excess?

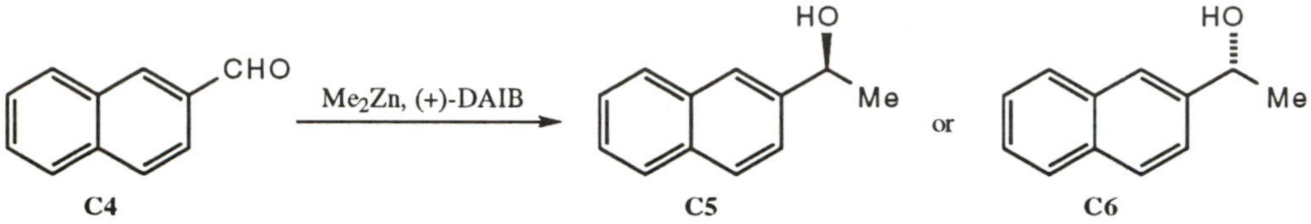

3.3 Reaction of **C7** with benzaldehyde could give in principle four possible products, but in practice one predominates. Which one will it be?

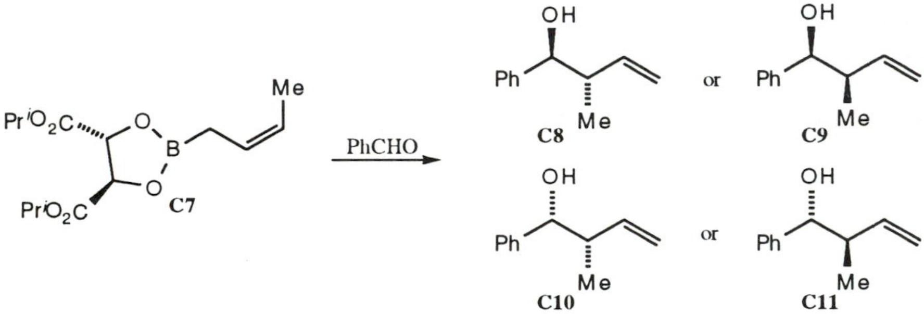

Suggestions for further reading

Additions to carbonyl groups are covered in detail in Chapters 5 and 6 of *Asymmetric Synthesis*, (ed. J. D. Morrison), Vol. 2, Academic Press, New York, 1983. For a discussion of the stereoelectronic requirements in additions to carbonyl groups, see E. L. Eliel, S. H. Wilen, and L. N. Mander, *Stereochemistry of Organic Compounds*, 876–888, Wiley, New York, 1994, and A. J. Kirby, *Stereoelectronic Effects* Oxford Chemistry Primer no. 36, 1996. For an introduction to catalytic enantioselective additions to aldehydes see R. Noyori and M. Kitamura, *Angew. Chem. Int. Ed. Engl.*, 1991, **30**, 49–69.

4 Reactions of enolates

4.1 The reaction

The reaction of enolates with electrophilic species, often referred to as α-substitution of carbonyl compounds, is a very important reaction in organic chemistry. One reason for this is that the carbonyl group can be converted easily into many other functional groups, indeed, the preceding chapter was concerned with just one such reaction of carbonyl compounds. The carbonyl group can even be removed completely by reduction to a methylene group (C=O to CH_2).

This means that we can use α-substitution to a carbonyl group to introduce a new substituent, and then convert to carbonyl group into another functional group (or into CH_2). This versatility makes the reaction particularly valuable in organic synthesis, and not surprisingly there has been much work carried out to devise methods to control the stereochemistry of substitution α to a carbonyl group.

Fig. 4.1 Enolate formation and alkylation

A typical general reaction is illustrated in Fig. 4.1, and as can be seen two steps are involved. The carbonyl compound **4.1** reacts with a base to give an enolate **4.2**, which then reacts with an electrophilic species (El) to give the product of α-substitution. No matter which approach is used to control the stereochemistry of the new stereocentre, control is usually required in both steps. We need to control both the geometry of the enolate and the stereochemistry of its reaction with an electrophile.

General considerations

The enolate has two faces which are either enantiotopic or diastereotopic depending on whether the enolate is prochiral or chiral. These are easily characterised as *re* or *si* by considering C-α of the enolate. The enolate possesses a double bond, so there is also the possibility of *E*/*Z* isomerism. These stereochemical relationships are shown in Fig. 4.2.

Descriptors (*E*)- and (*Z*)- are derived from the German for 'opposite' (*entgegen*) and 'together' (*zusammen*). The two substituents at each end of the double bond are allocated priorities as in the CIP convention. The isomer which has the two higher priority groups opposite is the (*E*)-isomer, and the other is the (*Z*)-isomer.

Stereoelectronic restrictions apply to the deprotonation step to form the enolate and its reaction with an electrophile. Deprotonation is favoured when the C–H bond which is broken is orthogonal to the plane of the carbonyl group. In such a conformation the σ C–H orbital is lined up with the π C=O

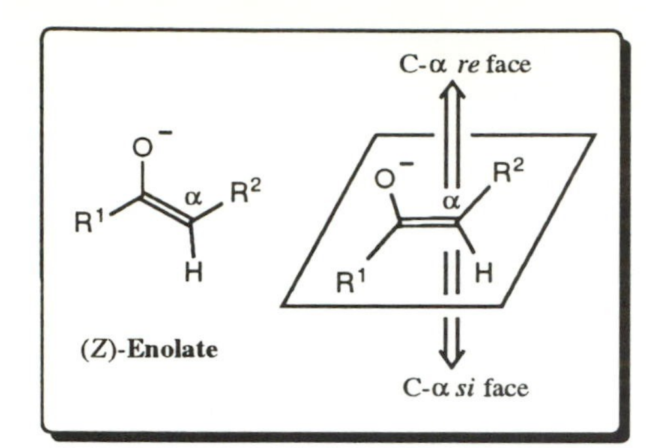

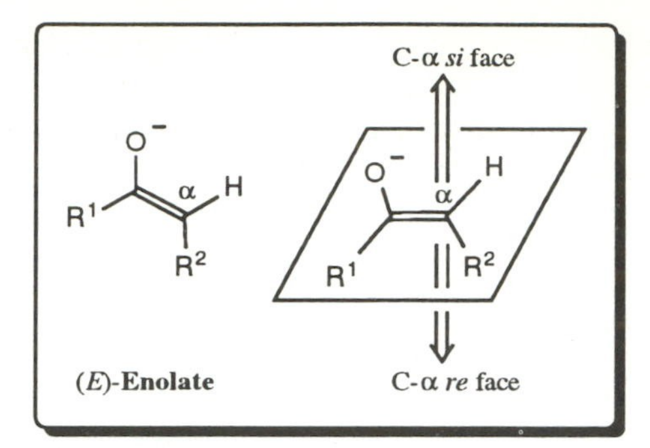

Fig. 4.2 Stereochemical descriptors for enolates

orbital (Fig. 4.3). This σ C–H orbital ultimately 'becomes' the p orbital at C-α of the enolate π bond, so it will be advantageous if it is already lined up with the p orbital of the carbonyl carbon. In addition, the starting conformation **4.4** is similar to the transition state **4.5**, so steric and electronic effects present in **4.4** should be reflected in **4.5**.

Fig. 4.3 Stereochemistry of deprotonation adjacent to a C=O group to form a (*Z*)-enolate

The oxygen atom of the enolate which originates from the C=O group is often arbitrarily given priority over the other substituent R1 irrespective of the rules of the CIP convention.

The corresponding (*E*)-enolate can be formed in a similar fashion, except that the relative positions of the atoms not involved directly in the reaction are different (Fig. 4.4).

Comparing the Newman projection **4.6** (Fig. 4.4) with **4.4** (Fig. 4.3), it is apparent that there is much greater steric interaction between R^1 and R^2 in the conformation which leads to the (*E*)-enolate. As stated above, such steric effects should also be reflected in the transition states **4.5** and **4.7**. This means that in cases where this interaction is large, we would expect the (*Z*)-enolate to be easier to form. Under kinetic control, the norm for most stereoselective reactions, the (*Z*)-enolate should predominate.

Reaction of the enolate with an electrophile will involve interaction of the

Fig. 4.4 Stereochemistry of deprotonation adjacent to a C=O group to form an (*E*)-enolate

HOMO of the enolate with the LUMO of the electrophile. A consequence of this is that the new bond to C-α will be formed more or less perpendicular to the plane of the carbonyl group, as illustrated (Fig. 4.5) for reaction with an alkyl halide (S_N2 reaction of the halide) from the C-α *re* face of a (*Z*)-enolate.

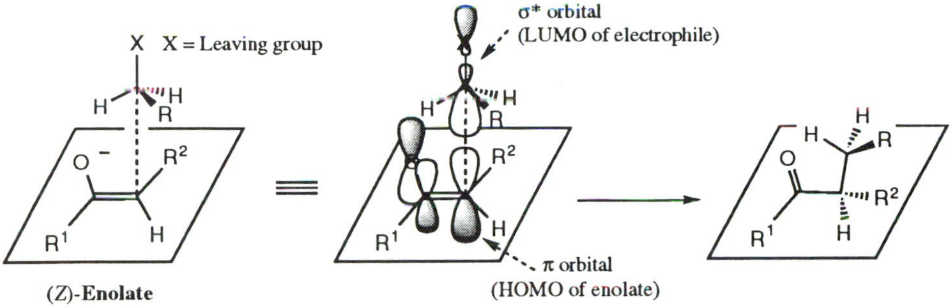

Fig. 4.5 Stereoelectronic requirement for addition of an electrophile to an enolate

In most of the examples which we will encounter, the group R^1 will be large, so we will be dealing with reactions of (*Z*)-enolates. Which face is favoured will depend on a number of things, usually the chirality of the group R^1. The largest section of this chapter is concerned with the reactions of such chiral enolates.

4.2 Reactions of chiral enolates

In some cases, the stereochemistry of enolate formation is unambiguous, in particular when the enolate forms part of a ring. The ring geometry allows only one enolate stereoisomer to be formed (Fig. 4.6).

The facial selectivity of alkylation of chiral enolates is largely governed by steric effects, so it is important to be aware of the stereochemistry and conformation of the enolate, and the preferred direction for bond formation.

In cases where the enolate is part of a ring, the conformation is rigid, and where large differences in steric hindrance on the two faces are present, the diastereoselectivity can be very high. The alkylation of **4.8** (Fig. 4.6) falls into this category. Being a *trans* decalin, the rings cannot flip, and the reaction takes place on a rigid framework. In particular, the methyl group in **4.8** and the enolate is effectively fixed with respect to the rest of the

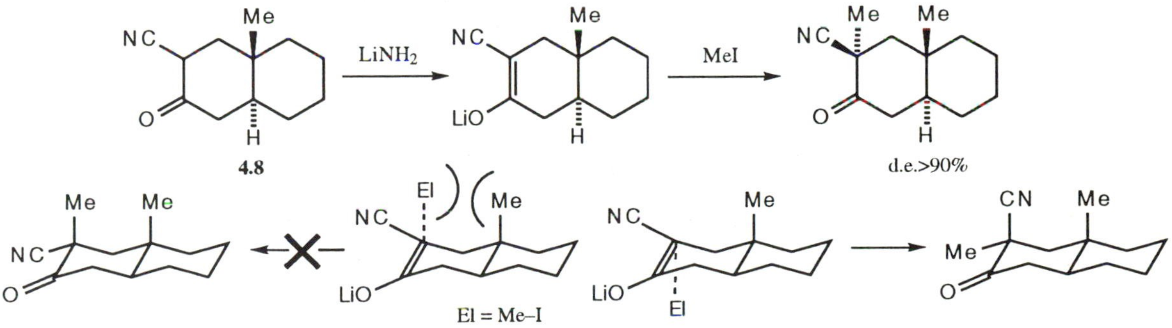

Fig. 4.6 Alkylation of a cyclic enolate

molecule. As the electrophile approaches from the same face of the enolate as this methyl group, large steric interactions build up. The corresponding transition state is disfavoured compared to attack on the other face, which is free of such severe steric interactions (Fig. 4.6).

Such reactions are not confined to polycyclic systems. The five membered lactone **4.9** can give only one enolate, which reacts with electrophiles from the less hindered face, opposite to the bulky *tert*-butyl group (Fig. 4.7).

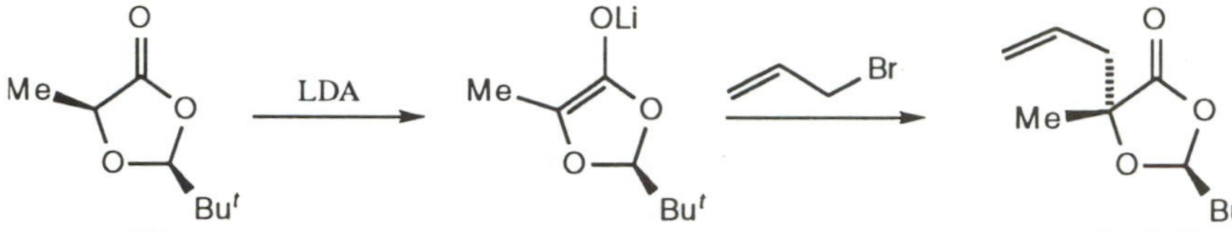

Fig. 4.7 Diastereoselective alkylation of a monocyclic enolate

α-Substitution of chiral acyclic enolates

The stereochemistry of the enolates which have so far been discussed has been determined by the structure of the ketone or ester, by virtue of it being part of a ring. When there is a stereocentre β to the carbonyl group in an acyclic ester, it is often possible to achieve high levels of diastereoselectivity. The general reaction is shown in Fig. 4.8. The simplest case is shown, in which two groups of very different steric demand are attached to the stereocentre. The most stable conformation is likely to be one in which the C–H bond of the stereocentre is more or less coplanar with the plane of the enolate.

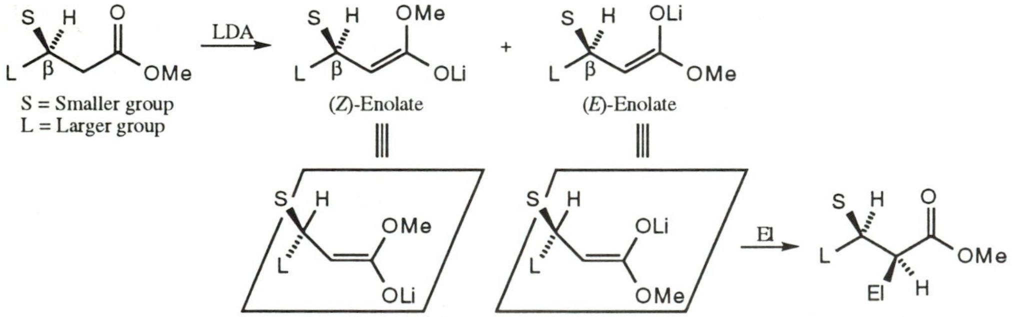

Fig. 4.8 Diastereoselective alkylation of a chiral acyclic enolate

This is an example of $A^{(1,3)}$ strain controlling the stereoselectivity in a reaction. This is discussed more fully in Chapter 6 (Fig. 6.7).

Electrophilic attack then takes place from the face of the enolate opposite to the larger group. The geometry of the enolate does not matter, as both OLi and OMe are both moderately sized groups. The key to this type of stereochemical control is the minimization of repulsive steric interactions between the groups attached to the stereocentre and the groups at the terminus of the enolate (OMe and OLi). If the difference in size between groups L and S is large enough, this type of reaction can take place with very high diastereoselectivity (Fig. 4.9).

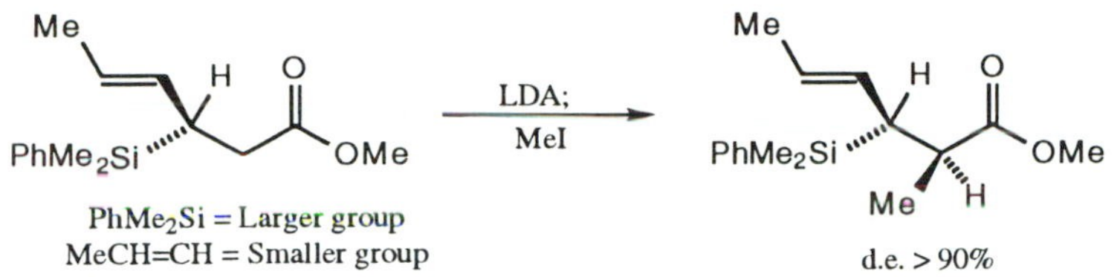

PhMe$_2$Si = Larger group
MeCH=CH = Smaller group

LDA; MeI

d.e. > 90%

Fig. 4.9 Diastereoselective alkylation of a chiral allylsilane

4.3 Chiral auxiliaries

Enolate geometry can be held rigidly in other ways. This, along with conformational control, is often a key aspect of efficient chiral auxiliaries for α-substitution. Initially we will consider systems derived from reaction of a chiral auxiliary with an activated carboxylic acid derivative.

Acyl derivatives

Oxazolidinone chiral auxiliaries **4.10** and **4.11** are particularly useful, and use strong chelation effects. They are easy to prepare from 1,2-amino alcohols, and to couple with a carbonyl compound (Fig. 4.10).

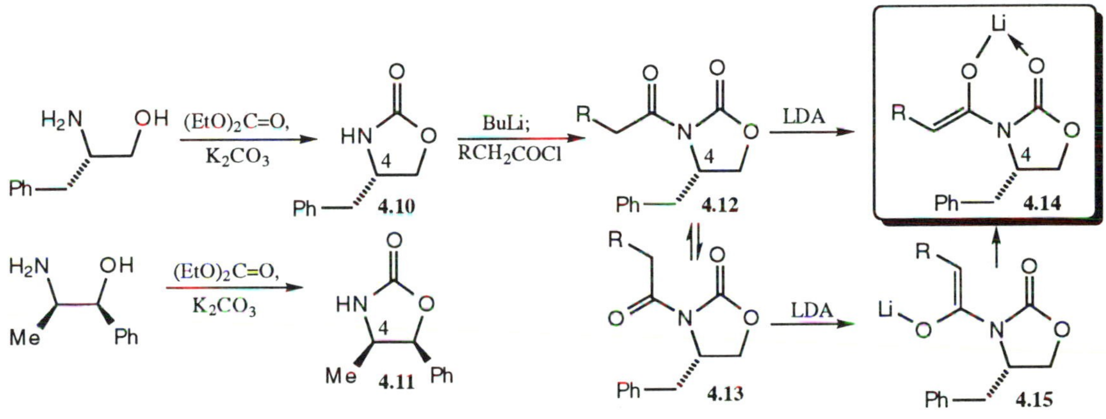

$(EtO)_2C{=}O$, K_2CO_3

BuLi; RCH_2COCl

LDA

Fig. 4.10 Preparation, coupling, and enolization of acyl oxazolidinones

Such acylated oxazolidinones can exist in two conformations, **4.12** and **4.13**, in which the nitrogen lone pair can delocalize into the carbonyl group of the acyl group. These will give the corresponding (*Z*)-enolate in two conformations **4.14** and **4.15**. Enolate **4.14** is much more stable due to chelation of the metal counter ion.

This chelation holds the enolate with respect to the chiral auxiliary, the oxazolidinone. In the reaction of such enolates, the electrophile is directed strongly by the stereochemistry at C-4. In **4.14** the lower face (C-α *re* face) is blocked by group at C-4, so the electrophile is directed to the upper (C-α *si*) face (Fig 4.9). Various metal enolates corresponding to **4.14** react with a range of electrophilic reagents in this way (Fig. 4.11).

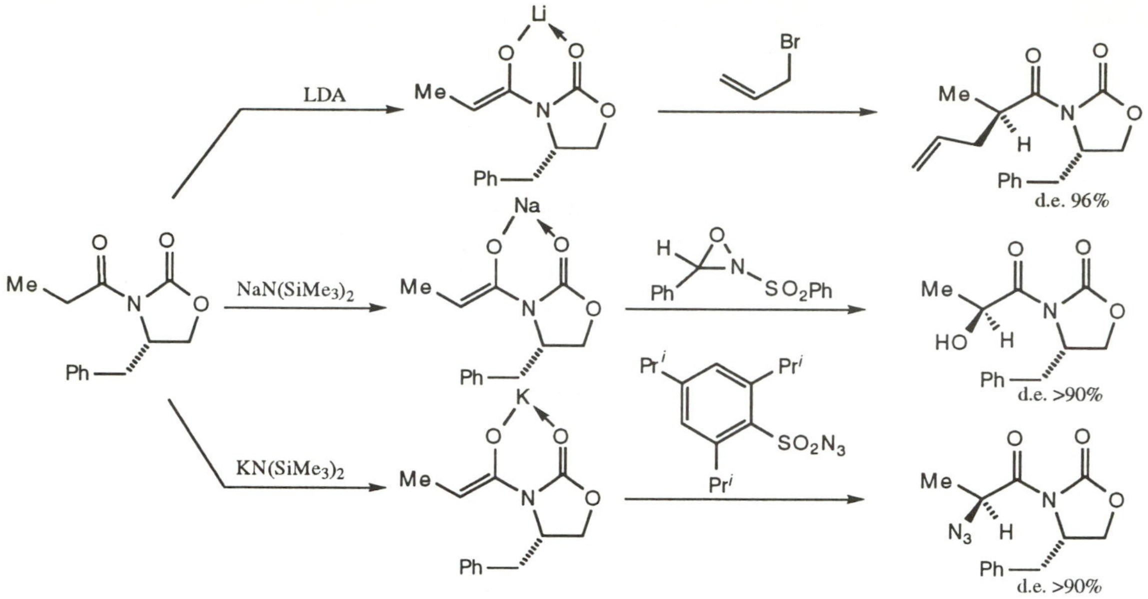

Fig. 4.11 Diastereoselective functionalization of acyl oxazolidinones

The oxazolidinone chiral auxiliaries discussed above are easy to remove from the product of α-substitution, being converted into carboxylic acids or esters, and primary alcohols (Fig. 4.12).

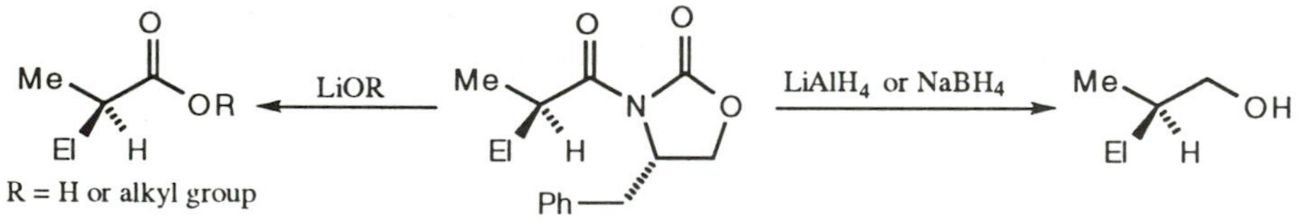

Fig. 4.12 Removal of oxazolidine chiral auxiliaries

Many other types of chiral auxiliaries have been developed for enolate α-substitution, two particularly interesting examples are shown in Fig. 4.13. In

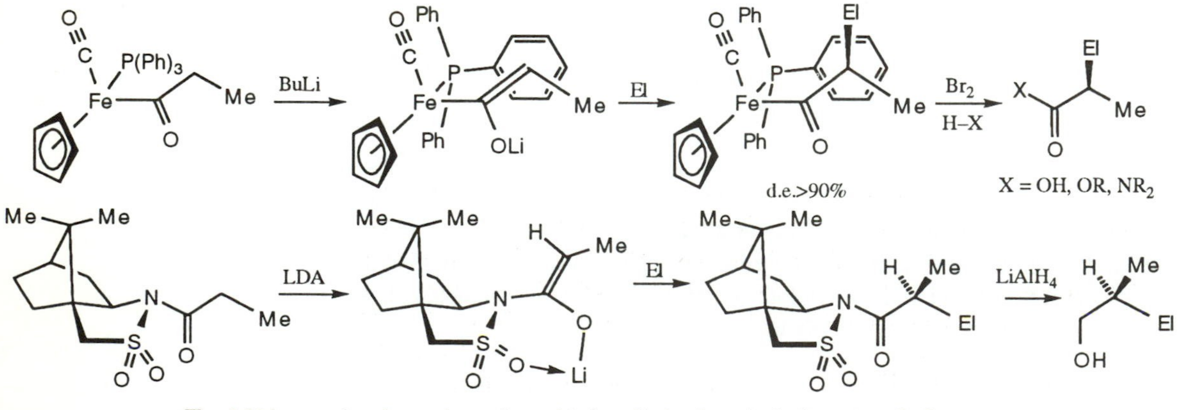

Fig. 4.13 Iron acyl and camphor sultam chiral auxiliaries for substitution α to a C=O group

both cases shown, a single enolate geometry is formed, it exists in one conformation, and part of the auxiliary effectively blocks one face of the enolate.

Both the chiral auxiliaries illustrated in Fig. 4.13 react with extremely high diastereoselectivity, and as with the oxazolidinones described above, are effectively chiral versions of a carboxylic acid or its derivatives. Although the method of chiral auxiliary removal is different for the iron acyl system, compared with the others, the overall effect is the same:- The formation of enantiomerically enriched acids, esters, amides, or alcohols. It is also possible to use chiral auxiliary methodology to achieve α-substitution of prochiral ketones, which is the subject of the next section.

α-Substitution of prochiral ketones and aldehydes

The most widely used type of chiral auxiliary for achieving α-substitution of prochiral ketones and aldehydes is based on conversion to an imine or a hydrazone, which carries a group capable of chelation. There are several such chiral auxiliaries, but we will concentrate on the most commonly encountered system, based on chiral hydrazines.

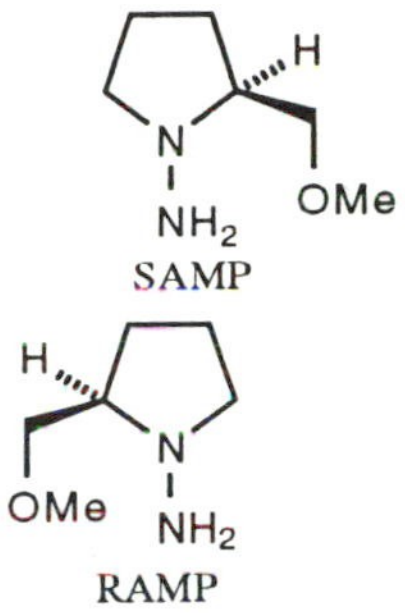

These enantiomeric auxiliaries, commonly known as RAMP and SAMP, are prepared from readily available enantiomerically pure starting materials, and react readily with ketones and aldehydes to give the corresponding hydrazones (Fig. 4.14).

Hydrazones derived from SAMP and RAMP react with strong base, usually butyllithium, to give a species analogous to an enolate. This 'enolate' is thought to possess a rigid structure, which results in highly diastereoselective reaction with alkylating agents. The chiral auxiliary cannot be removed easily by simple hydrolysis. Ozonolysis of the C=N bond is efficient, and provides the corresponding carbonyl compound and the *N*-nitroso derivative, which is an intermediate in the synthesis of the chiral auxiliary (Fig. 4.14).

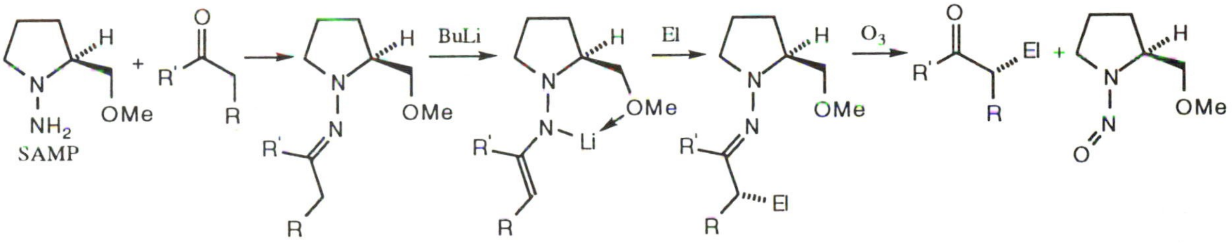

SAMP

RAMP

Fig. 4.14 Use of a SAMP hydrazone for α substitution

The electrophiles which have been used include a wide range of alkyl halides, and the diastereoselectivity in the alkylation step can be very high, typically in the range 80 to ≥95 per cent. A typical example of this approach is shown in Fig. 4.15, and demonstrates its simplicity and effectiveness. The SAMP hydrazone of *n*-butanal, following conversion to the 'enolate' with LDA, reacts with the alkyl iodide **4.16** in high d.e. After removal of the chiral auxiliary, ketone **4.17** was used in a total synthesis of an ionophore antibiotic X-14547A.

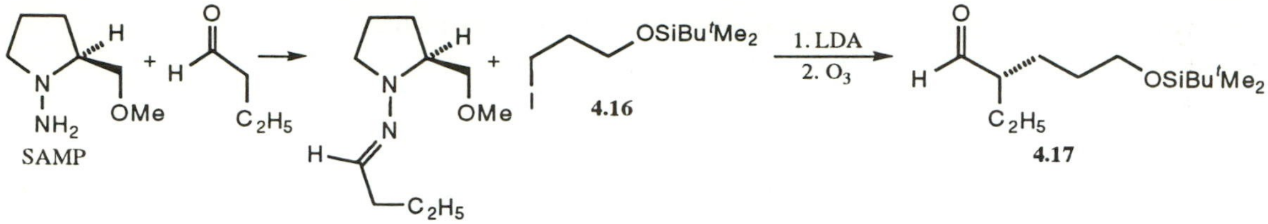

Fig. 4.15 Preparation of an intermediate in the total synthesis of X-14547A

Although most α-substitutions of prochiral ketones and aldehydes is based on chiral auxiliary methodology, it is possible to use both chiral reagents and chiral catalysts. The final two sections of this chapter are concerned with these approaches.

4.4 Chiral reagents

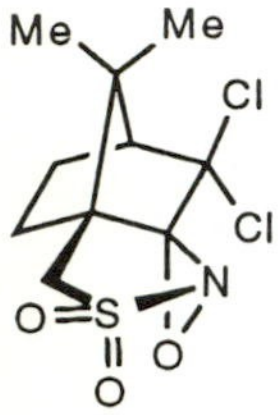

α-Substitution of a ketone involves two other reagents, the electrophile and the base. In principle enantioselective α-substitution involving a prochiral enolate should be possible with either a chiral electrophile or base. It is possible to achieve such reactions, and although neither have achieved the generality of chiral auxiliary methodology, both approaches show promise.

One of the most successful groups of chiral electrophiles is based on oxaziridines such as **4.18**. These derive ultimately from camphor, and both enantiomers are available. Oxaziridine **4.18** reacts with ketone enolates to give the corresponding α-hydroxy ketone, usually in high e.e.

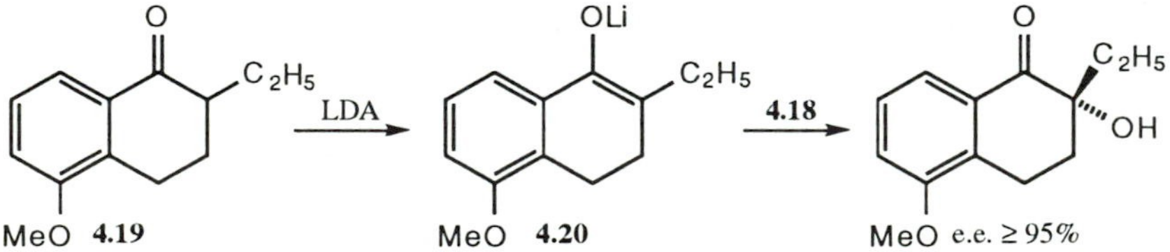

Fig. 4.16 Enantioselective α hydroxylation

In the example shown in Fig. 4.16, a racemic ketone **4.19** was converted into a prochiral enolate **4.20**, and the enantioselectivity originated from the chirality of the electrophile. It is also possible to take a prochiral ketone, and generate from it a chiral enolate. Reaction of the chiral enolate with an achiral electrophile should result in an enantioselective reaction.

This has been achieved by using a chiral base to 'desymmetrize' prochiral ketones. Usually it is necessary to use the chiral base in conjunction with trimethylsilyl chloride to trap the enolate as it is formed. The resulting

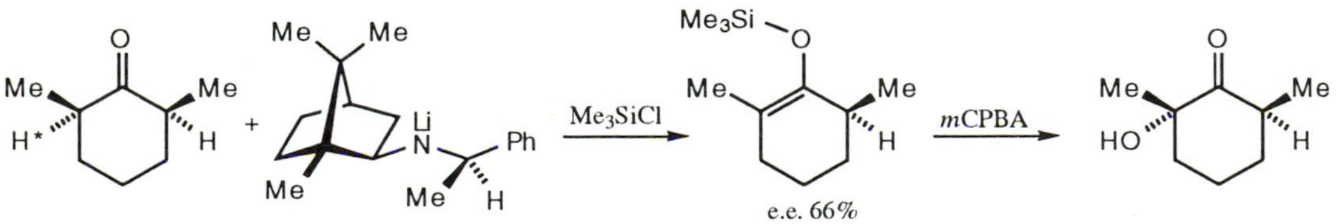

Fig. 4.17 Use of a chiral base in α hydroxylation

trimethylsilyl enol ethers can be formed in good to high e.e., and used in subsequent reactions. A simple example is given in Fig. 4.17.

4.5 Enantioselective catalysis

Enantioselective alkylation of enolates is rather difficult to achieve, but it is possible to carry out some types of alkylation with remarkably high enantioselectivity. Some of the best asymmetric catalysts are quaternary ammonium salts derived from readily available alkaloids such as quinine and cinchonidine.

4.21

The quaternary ammonium catalyst (QAC) **4.21** is derived from cinchonidine, and acts as a highly efficient enantioselective phase transfer catalyst for the alkylation of esters such as **4.22**. This is an important process as the products can easily be converted into the corresponding amino acid esters, and the reaction works well for a wide range of alkylating agents.

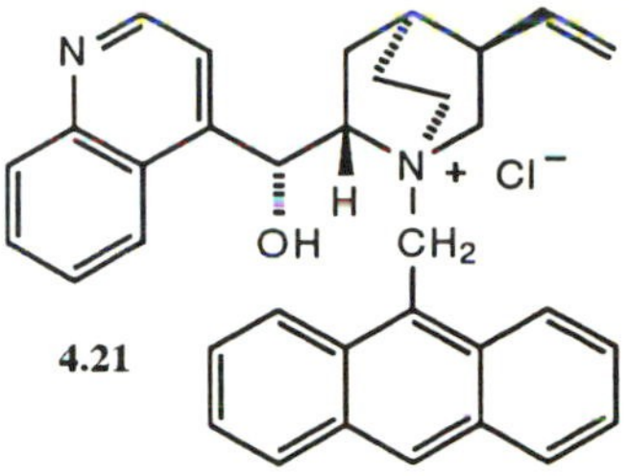

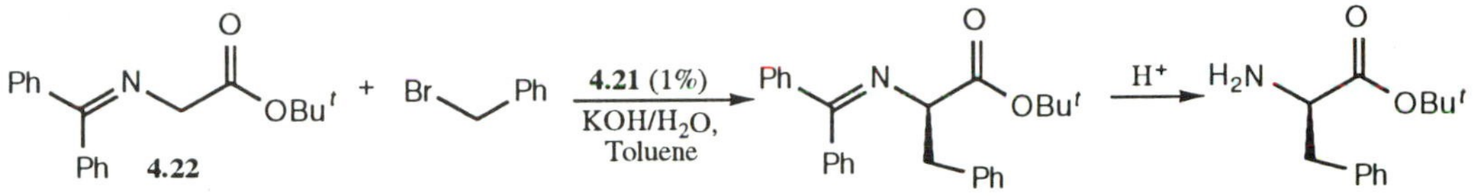

Fig. 4.18 Enantioselective alkylation using phase transfer catalysis

Phase transfer catalysts such as **4.21** work by 'transporting' hydroxide into the organic phase, which contains the substrate and the alkylating agent, as an ion pair (Fig. 4.19). In an organic solvent, hydroxide is a powerful base and deprotonates the carbonyl compound, forming a new ion pair in which the anionic species is the enolate. This tight ion pair is chiral, and so the faces of the enolate are effectively diastereotopic, so stereoselective alkylation is possible. After alkylation, the catalyst is 'free' to transport another hydroxide ion into the organic phase, so continuing the catalytic cycle.

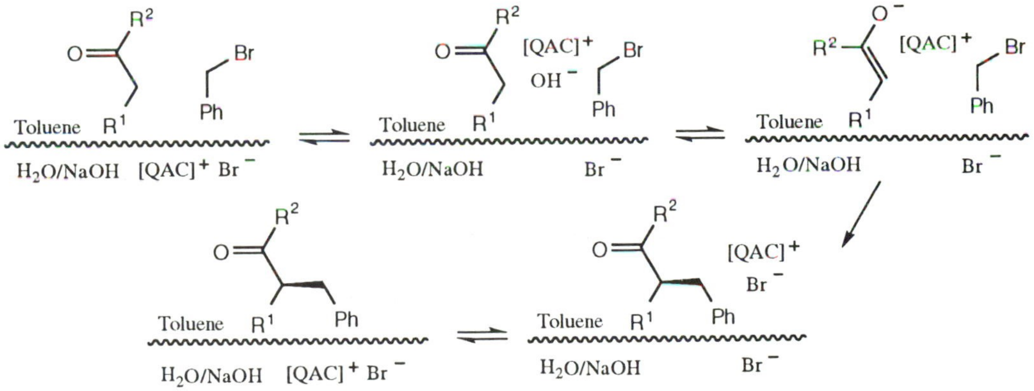

Fig. 4.19 Catalytic cycle for enantioselective phase transfer alkylation

4.6 Problems

4.1 Which diastereoisomer would you expect to predominate in the alkylation of **D1**?

Et O O N O Me Ph **D1** — LDA; MeI → Et O O Me N O Me Ph **D2** or Et O O Me N O Me Ph **D3**

4.2 Account for the stereoselectivity in the alkylation of **D4**, which gives **D5** in very high d.e.

Me CO_2Me $SiPh_3$ **D4** — LDA; MeI → Me Me CO_2Me $SiPh_3$ **D5**

Suggestions for further reading

The stereoselective alkylation of metal enolates and related reactions are discussed at length in Chapters 1, 3, and 4 of *Asymmetric Synthesis*, (ed. J. D. Morrison), Vol. 3, Academic Press, New York, 1984. Enolate formation and reactions of chiral enolates with electrophiles are covered in E. L. Eliel, S. H. Wilen, and L. N. Mander, *Stereochemistry of Organic Compounds*, 899–905, Wiley, New York, 1994. For enantioselective aspects of phase transfer catalysis in enolate alkylation see Chapter 8 in Catalytic Asymmetric Synthesis, (ed. I. Ojima), VCH, New York, 1993.

5 Aldol reactions

5.1 The reaction

The aldol reaction (Fig. 5.1) occupies a particularly important position in organic synthesis. It is a very general method for forming a new C–C bond, and at the same time two new stereocentres. Moreover, the aldol unit can be found in the structures of numerous natural products, many of which possess valuable and interesting biological activities, making them prime targets for the attentions of synthetic chemists. As a result much attention has been focused on the reaction, with some spectacular results. In this chapter we can only scratch the surface of this fascinating and invaluable synthetic reaction.

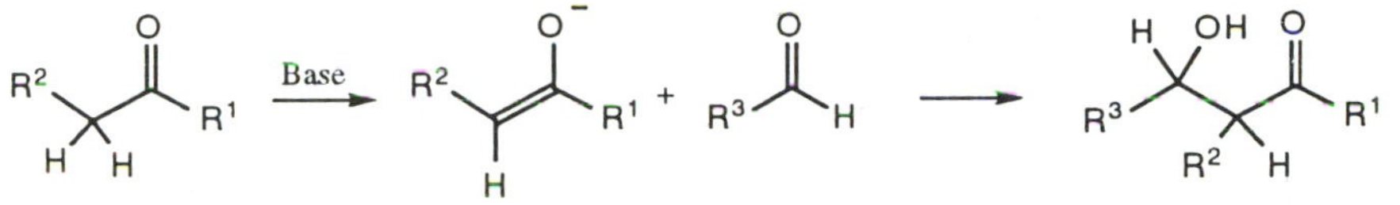

Fig. 5.1 The aldol reaction

The stereochemical aspects of this type of reaction are discussed in Chapter 1.

Two new stereocentres are created from two prochiral carbon atoms, so there are four possible product diastereoisomers (Fig. 5.2). In the simplest case, when neither enolate nor aldehyde are chiral (and there is no chiral catalyst or reagent present), the products are two racemic diastereoisomers. These are usually referred to as the *syn*- and the *anti*-aldol products (Fig. 5.2). The *syn*- product has the substituents R^2 and OH on the same face of the molecule, when the 'main chain' is drawn in an extended fashion. The *anti*-product has these substituents on opposite faces.

Many aldol reactions take place through a highly ordered transition state, which appears to possess a predictable stereochemistry. This is often called

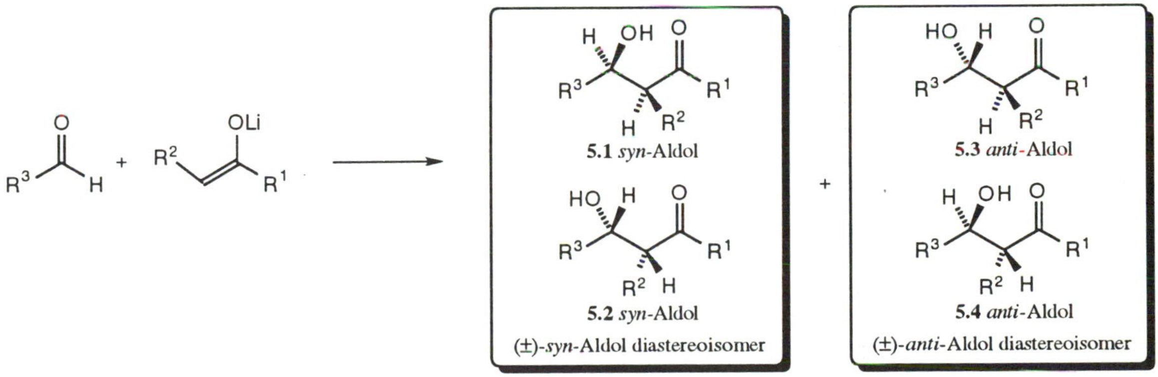

5.1 *syn*-Aldol

5.2 *syn*-Aldol

(±)-*syn*-Aldol diastereoisomer

5.3 *anti*-Aldol

5.4 *anti*-Aldol

(±)-*anti*-Aldol diastereoisomer

Fig. 5.2 Stereochemical descriptors for diastereoisomeric aldol products

the Zimmerman–Traxler transition state (Fig. 5.3). In this, the counter ion of the enolate, usually a metal, is bonded to the oxygen of the enolate and co-ordinated to the oxygen of the aldehyde. Following this 'complexation' the bond forming and bond breaking take place.

The Zimmerman–Traxler transition state is analogous to that of the Claisen and Cope rearrangement, which was discussed in some detail Chapter 2 (Section 2.3). As in these rearrangements, the aldol transition state is six membered, and can usefully be considered to be similar to a cyclohexane. In spite of the fact that it must be very far from being a cyclohexane, this analogy is very useful in understanding, and to some extent predicting, the stereochemical outcome of many aldol reactions.

5.2 Stereochemical aspects of aldol reactions

If we regard the Zimmerman–Traxler transition state as possessing an overall geometry similar to a cyclohexane (Fig. 5.3), then we can make some predictions about the likely stereochemical outcome of a given reaction.

(*E*)-Enolate

Zimmerman–Traxler transition state

re Face of enolate adds to *re* face of aldehyde

5.3
anti-Aldol

Fig. 5.3 Chair-like transition state accounts for the correlation of (*E*)-enolates with *anti*-aldols

In this discussion we will assume that O–M (e.g. OLi) in an enolate is always higher priority than R^1.

The most stable arrangement is likely to be that in which all the substituents are equatorial. This is possible if we start from an (*E*)-enolate. It will always be possible for the larger R^3 group of the aldehyde to be equatorial, as there are no constraints on the aldehyde other than the C=O group becomes part of the six membered ring. A typical transition state for reaction of an (*E*)-enolate is illustrated in Fig. 5.3. As can be seen, this results in an *anti*-aldol.

The same (*E*)-enolate can also react via a transition state enantiomeric to that shown in Fig. 5.3. This gives the enantiomeric product, which must be formed at the same rate, leading to an equal amount of the enantiomeric *anti*-aldol (Fig. 5.4) (enolate and aldehyde are achiral, and no other chiral component is involved).

At first sight it might be thought that the two transition states leading to **5.3** and **5.4** are related by a 'ring flipping' of the chair-like transition state. This is not the case. Such a ring flipping exchanges equatorial groups and axial groups. Ring flipping of either transition state would place R^2 and R^3

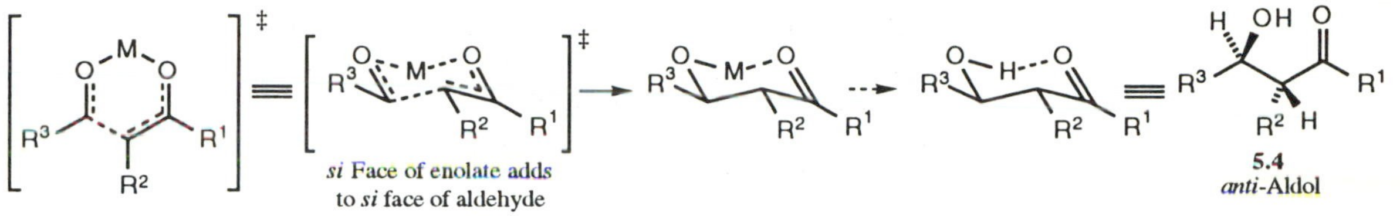

si Face of enolate adds to *si* face of aldehyde

5.4 *anti*-Aldol

Fig. 5.4 Reaction from the *si* face of the aldehyde gives the enantiomeric *anti*-aldol (see Fig. 5.3)

in axial positions, which clearly is not the case. What has happened is that the reactants have approached and reacted from the opposite face of each. There is no equilibrium between the two, as almost all highly stereoselective aldol reactions are under strict kinetic control.

Given the above discussion, it follows that if we change the geometry of the enolate, then the stereochemistry of the aldol should be different. The enolate geometry is fixed under the reaction conditions, so by analogy with the above, the (*Z*)-enolate should result in substituent R^2 occupying an axial position (assuming the aldehyde substituent R^3 can always be equatorial). Thus, aldol reactions of (*Z*)-enolates generally give a *syn*-aldol (Fig. 5.5).

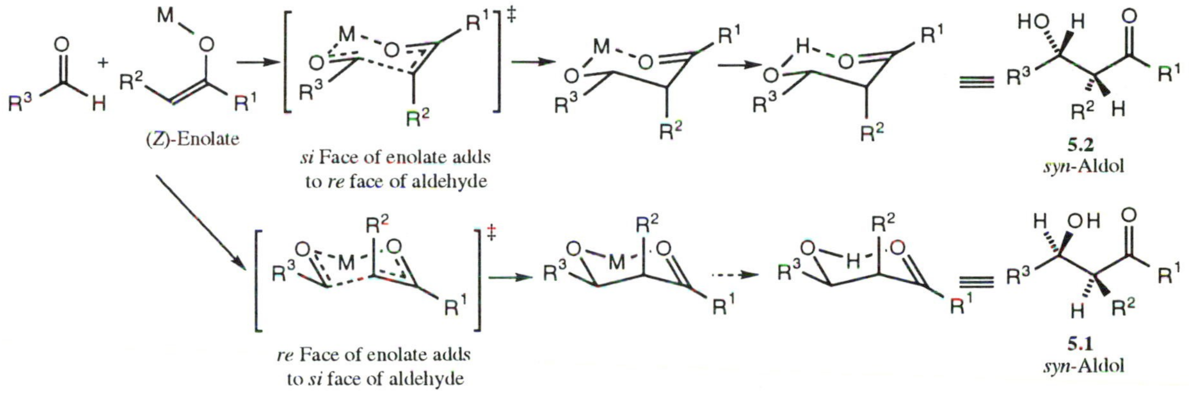

(*Z*)-Enolate

si Face of enolate adds to *re* face of aldehyde

5.2 *syn*-Aldol

re Face of enolate adds to *si* face of aldehyde

5.1 *syn*-Aldol

Fig. 5.5 Correlation of (*Z*)-enolates with *syn*-aldol

From the foregoing arguments, it would seem that we simply have to control the enolate geometry and high levels of diastereoselectivity will follow. In practice this is not always the case, the stereoselectivity can depend on many aspects of the reaction, including the nature of the metal ion, the structure of the enolate, and the reaction conditions. Nevertheless, methods have been developed which are reasonably predictable and highly stereoselective.

5.3 Control of enolate geometry

Some important aspects of enolate formation have been discussed previously in Chapter 4 (Section 4.1). This is actually a rather complicated subject, and this section can only cover some of the more important trends (Fig. 5.6). If the carbonyl group lies within a ring, then the enolate geometry will be

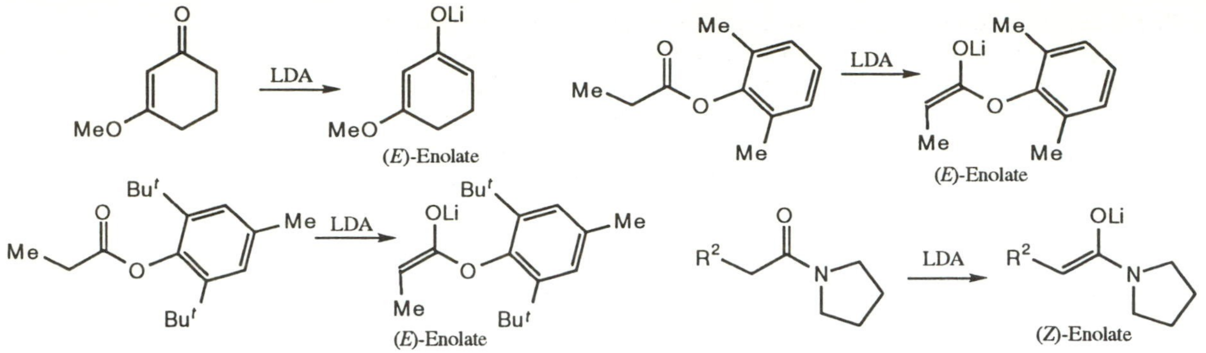

Fig. 5.6 Stereochemistry of enolates from some typical C=O compounds

fixed, provided the ring is not very large. The most common ring sizes (4–7) can only easily accommodate (*E*)-enolates.

If the carbonyl group is not part of a ring, then both enolate geometries are possible, and it is possible to favour one or the other by judicious choice of reagent, reaction conditions, and structure of the carbonyl compound.

Esters tend to give (*E*)-enolates on deprotonation with LDA, and esters of bulky phenols are particularly useful in aldol reactions. In contrast, bulky amides give (*Z*)-enolates, and although the lithium enolates of amides are not particularly useful in aldol reactions, other metal (*Z*)-enolates of amides and related systems can be very effective.

Enolization of acyclic ketones with bases such as LDA is rather more complicated, and not appropriate for detailed coverage in this book. However, the stereoselective formation of boron enolates of acyclic ketones is usually more straightforward. Stereoselective formation of boron enolates of esters, amides, and related derivatives is also rather predictable. Moreover, the aldol reactions of such enolates often take place with predictable stereoselectivity, as there tends to be a much stronger correlation between enolate geometry and aldol stereochemistry for boron enolates.

Boron enolates are usually generated by reaction of the carbonyl compound with a dialkylboron reagent possessing a leaving group, and a weak tertiary

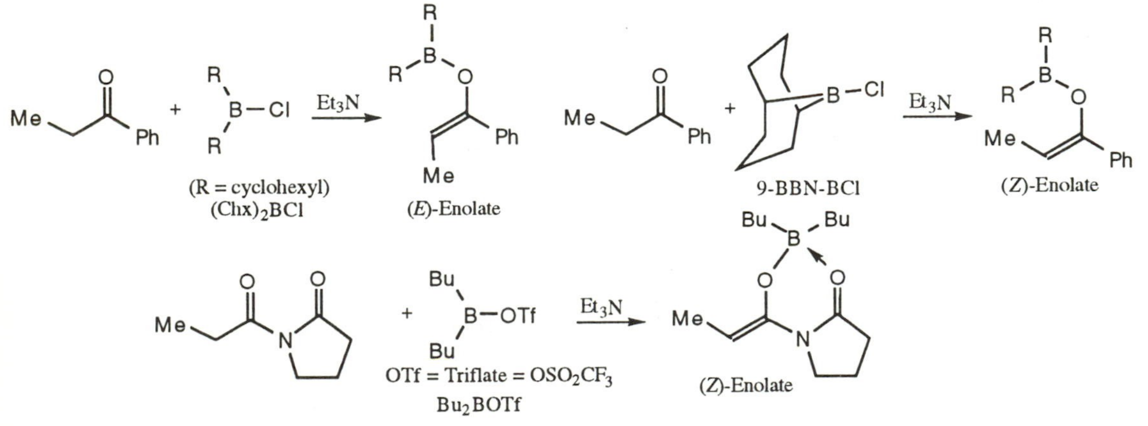

Fig. 5.7 Stereoselective formation of boron enolates

amine base. The boron reagent is a Lewis acid and co-ordinates to the carbonyl group, activating the α protons towards removal with a weak base. The geometry of the resulting enolate depends on the nature of the carbonyl compound, the leaving group, and to some extent on the base. A thorough discussion is beyond the scope of this book, but some representative examples are given in Fig. 5.7. In all cases shown, the selectivity is > 95:5 for the enolate shown.

5.4 Diastereoselective reactions involving prochiral substrates

This is the type of reaction which we used in Section 5.1 to demonstrate some of the stereochemical aspects of aldol reactions. It necessarily leads to racemic products, but is useful nevertheless as high levels of diastereoselectivity can be achieved in the best cases. Both lithium and boron enolates can be used with the appropriate substrate (Fig. 5.8).

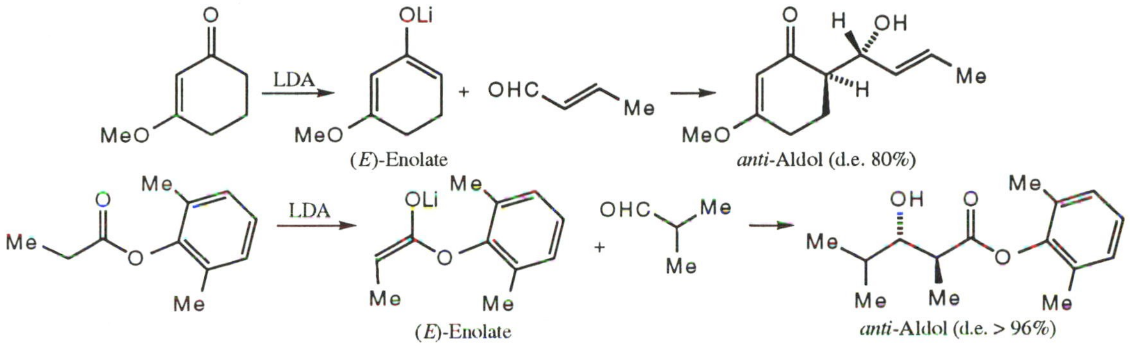

Fig. 5.8 Examples of *anti*-aldol reactions using lithium enolates

Boron enolates generally exhibit high diastereoselectivity, and good correlation of enolate geometry with the aldol stereochemistry (Fig. 5.9). One reason for this is undoubtedly that the maximum co-ordination number of boron is four. This means that once boron enolates such as those shown in Fig. 5.9 have co-ordinated to the aldehyde (see Fig. 5.3), then no more ligands can be accommodated around the boron.

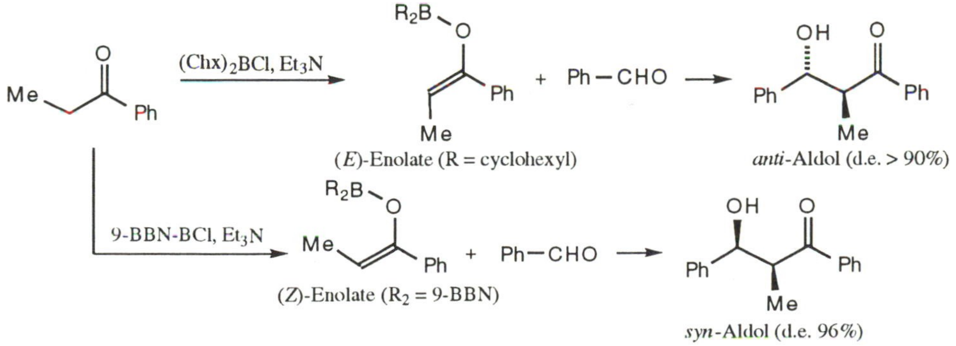

Fig. 5.9 Examples of aldol reactions using boron enolates

5.5 Diastereoselective reactions involving chiral enolates

Both aldehyde and enolate involved in an aldol reaction can be chiral, but we will limit our discussion to aldol reactions involving chiral enolates. In general these are likely to be used as single enantiomers. As usual, we can but scratch the surface of this important topic, but we will cover the most important aspects by considering examples of some of the more important types of reactions.

Chiral ketones

Ketones which possess an oxygen substituent α or β to the carbonyl group have been studied most intensively, as the corresponding boron enolates often react with high diastereoselectivity. Ketone **5.5** and its enantiomer are readily available, and following conversion to the (*Z*)-enolates, undergo highly diastereoselective aldol reactions with a wide range of aldehydes (Fig. 5.10).

The original stereocentre and its associated substituents can be excised by removal of the silyl ether (with HF) and oxidation of the resulting α hydroxy acid (with $NaIO_4$) as shown for the conversion of **5.6** into **5.7** (Fig. 5.10). Effectively, the 'right hand side' of ketone **5.5** is used as a chiral auxiliary. More 'normal' types of chiral auxiliaries are discussed later.

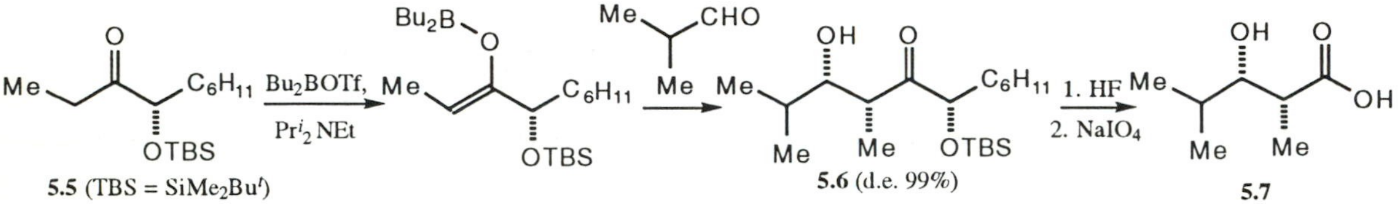

5.5 (TBS = $SiMe_2Bu^t$) — Bu_2BOTf, Pr^i_2NEt → boron enolate; + Me(Me)CHCHO → 5.6 (d.e. 99%) — 1. HF, 2. $NaIO_4$ → 5.7

Fig. 5.10 Aldol reactions involving chiral ketones with a single stereocentre

If the ketone has an oxygen substituent β to the carbonyl group, then the ketone itself could have been derived from an asymmetric aldol reaction. The ability to carry out a subsequent aldol reaction would then amount to an iterative process, in which the stereochemistry of the previous aldol reaction controls that of the next aldol reaction. Stereochemical control is possible in reactions involving such ketones, and an example is provided in Fig. 5.11.

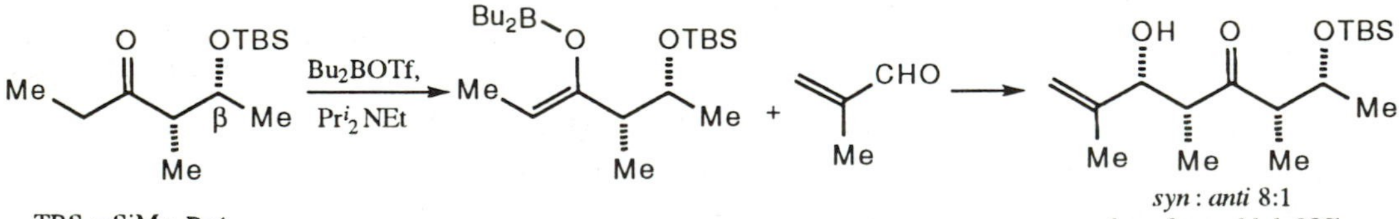

TBS = $SiMe_2Bu^t$ — Bu_2BOTf, Pr^i_2NEt → boron enolate + methacrolein → aldol; *syn* : *anti* 8:1; d.e. of *syn*-aldol 90%

Fig. 5.11 Aldol reaction involving chiral ketones with a two adjacent stereocentres

In both examples provided above, the enolate geometry (*Z*) results in a *syn*-aldol product. However, as there is a pre-existing stereocentre in the ketone, the two possible *syn*-aldol products are diastereoisomers. The major product arises from reaction from the *si* face of the enolate and the *re* face of

si Face of enolate adds to *re* face of aldehyde

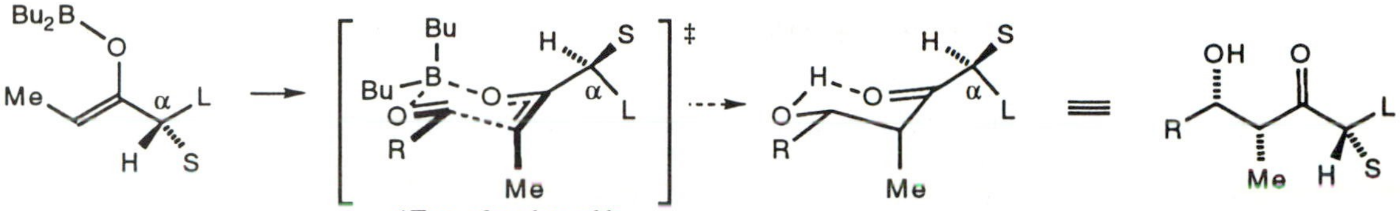

Fig. 5.12 Transition state model for the aldol reaction of an enolate with an adjacent stereocentre

the aldehyde (see Fig. 5.5). These reactions are consistent with a transition state model similar to that illustrated in Fig. 5.12.

Chiral auxiliaries

There are many chiral auxiliaries available for controlling the stereochemistry of aldol reactions, two of which are shown in Fig. 5.13. Rather than attempt a comprehensive discussion of all chiral auxiliaries which have been used for the aldol reaction, we will restrict ourselves mainly to the oxazolidinones.

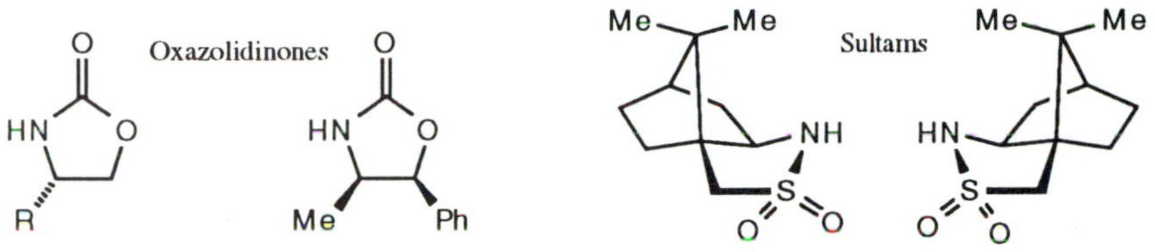

Oxazolidinones Sultams

Fig. 5.13 Chiral auxiliaries for aldol reactions

We have encountered such oxazolidinones before, as they are particularly important chiral auxiliaries for enolate α-substitution (Chapter 4, Section 4.3). They are also very effective chiral auxiliaries for the aldol reaction. In particular, they are easily acylated and the acylated products form boron enolates which possess exclusively the (*Z*) geometry. These enolates react with high diastereoselectivity to give the corresponding *syn*-aldol products (Fig. 5.14). The diastereoselectivity is usually very high, and predictable in the sense shown. The auxiliary can be removed with a variety of reagents (see Chapter 4, Fig. 4.12). As shown in Fig. 5.14, the either enantiomeric product can be prepared depending on which oxazolidinone is used.

The facial selectivity of the enolate implicit in the reactions shown in Fig. 5.14 is consistent with a transition state model not dissimilar to that encountered previously in the reactions of chiral ketones (Fig. 5.12). Although the enolates themselves involve co-ordination of the boron, this must be broken for the boron to co-ordinate to the aldehyde, a requirement in

An aldol reaction with the other oxazolidinone is shown in Fig. 5.15.

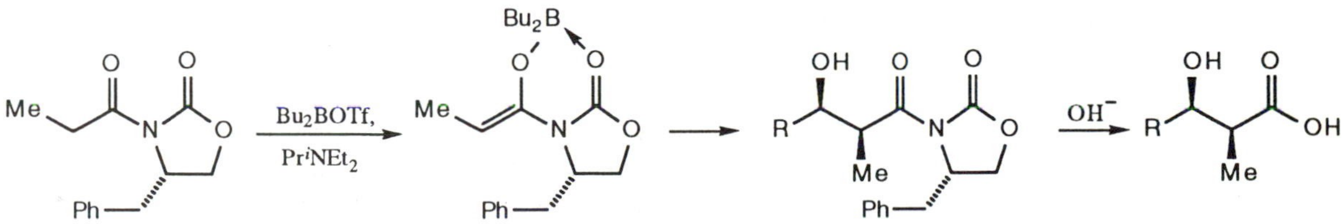

Bu_2BOTf, Pr^iNEt_2 OH^-

Fig. 5.14 Oxazolidinones as chiral auxiliaries for aldol reactions

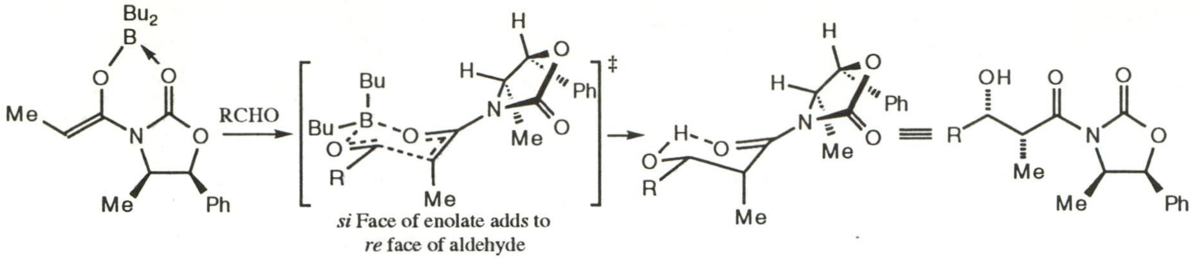

Fig. 5.15 Transition state model for aldol reactions involving the boron enolates of acyl oxazolidinones

these aldol reactions. The chiral auxiliary is then free to adjust its position so that electronic and steric repulsions are minimized. These considerations form the basis for the transition state model illustrated in Fig. 5.15.

5.6 Chiral reagents and catalysts

A number of chiral reagents which have been developed for the aldol reaction, many of which rely upon the use of chiral boron species, such as **5.8** (Fig. 5.16), the reactions of which can be controlled to give either diastereoisomer of the product in high e.e. (83-98 per cent).

Reagents related to **5.8** have been used for addition to carbonyl groups, substitution α to carbonyl groups, Diels–Alder cycloadditions, and rearrangement reactions.

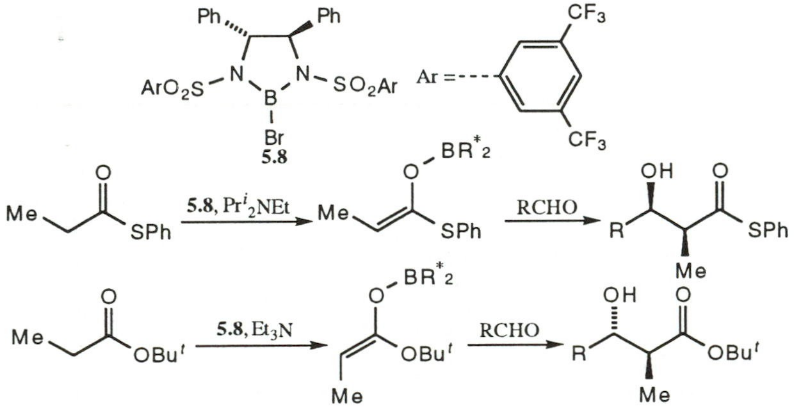

Fig. 5.16 Aldol reaction using a chiral reagent

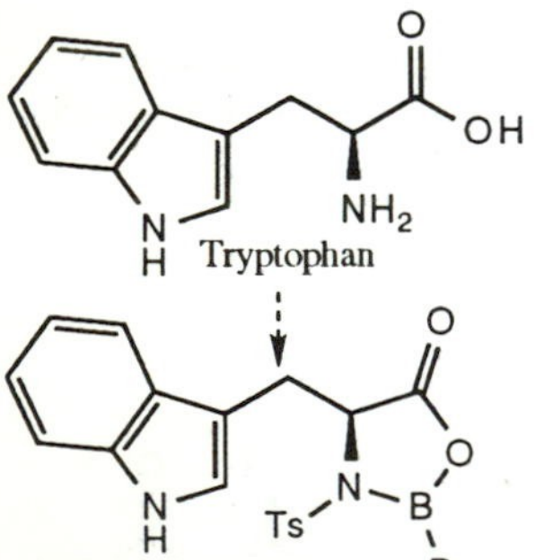

There are also a number of chiral catalysts which are capable of achieving high levels of enantioselectivity in some types of aldol reactions. One such catalyst **5.9** in Fig. 5.17. It is derived from the amino acid tryptophan, and catalyses aldol reactions of silyl enol ethers such as **5.10**.

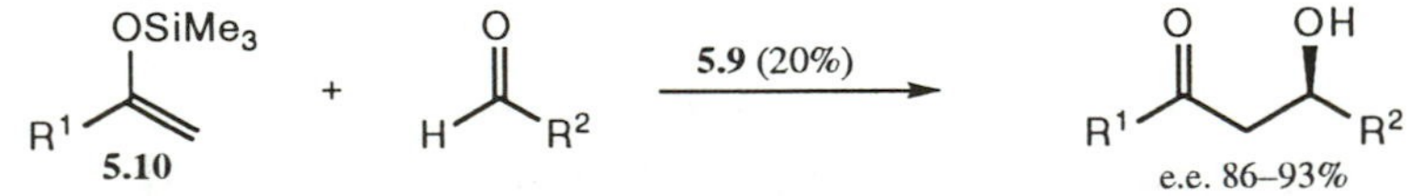

Tryptophan

5.9 Ts = *p*-$MeC_6H_4SO_2$

Fig. 5.17 Aldol reaction using a chiral catalyst

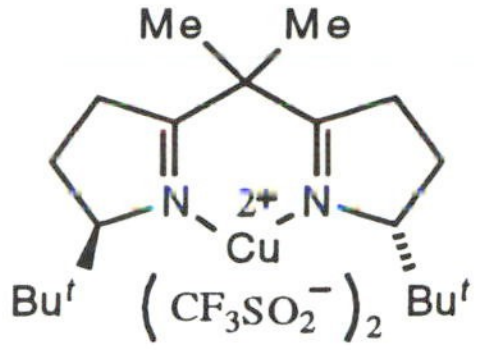

Catalyst **5.11** is capable of achieving remarkable selectivity in the reactions of silyl enol ethers such as **5.12** (Fig. 5.18). In the example shown, the catalyst is capable of bringing about an aldol reaction with the unsymmetrical diketone **5.13** in high d.e. and e.e. Moreover, reaction takes place at the less hindered of the two carbonyl groups to give **5.14** selectively. The steric demands of ethyl and methyl groups are not that different, and the observed regioselectivity is quite remarkable.

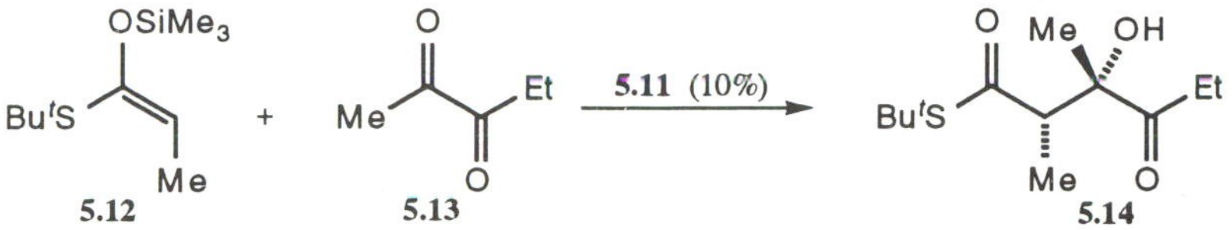

Me Me
N 2+ N
Cu
Bu^t $(CF_3SO_2^-)_2$ Bu^t
5.11

OSiMe$_3$ Bu^{t}S Me **5.12** + Me O Et O **5.13** —**5.11** (10%)→ Bu^{t}S O Me OH Et Me O **5.14**

Regioselectivity 98:2; d.e. 86%; e.e. 97%

Fig. 5.18 Aldol reaction of as ketone using a chiral catalyst

5.7 Problems

5.1 Assuming a Zimmerman–Traxler transition state, what would you expect to be the major product of the aldol reaction of **E1** with 2,2-dimethylpropanal?

O **E1** —LDA; Bu^{t}CHO→ **E2**

5.2 The chelated boron enolate **E3** undergoes aldol reaction through transition state **E4**. Why is the chelation present in **E3** absent in **E4**, and why does the chiral auxiliary adopt the conformation shown?

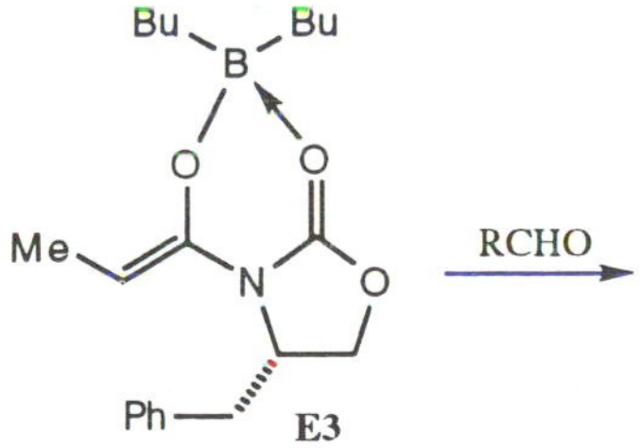

Bu Bu B O O Me N O Ph **E3** —RCHO→ Bu Me O B O R Bu Ph N O O **E4**

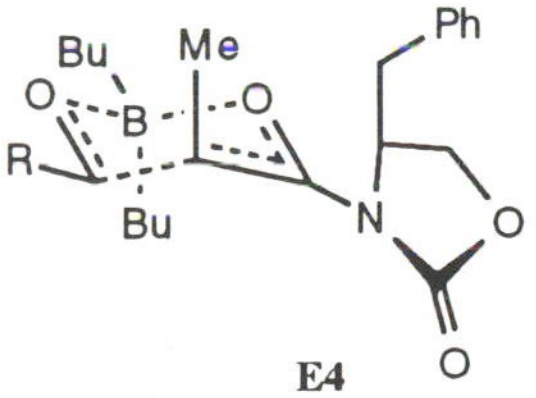

Suggestions for further reading

A thorough analysis of the aldol reaction is to be found in Chapter 2 of *Asymmetric Synthesis*, (ed. J. D. Morrison), Vol. 3, Academic Press, New York, 1984. For a more recent review, which encompasses literature up to 1993, see A. S. Franklin and I. Paterson, *Contemporary Organic Synthesis*, 1994, **1** 317–338.

6 Additions to C–C double bonds

6.1 The reactions

There are two particularly important types of additions to C–C double bonds. The reaction can be concerted, in which a new bond is formed to each end of the double bond more or less simultaneously. At the other extreme, the two bond forming events are effectively independent of each other, except that for mechanistic reasons one must precede the other. In this chapter we will consider a selection of some of the more important types of additions, including hydroboration, Diels–Alder cycloaddition, electrophilic addition, and conjugate addition. Some basic aspects of the mechanism and stereoelectronic requirements of these reactions have been covered previously in Chapter 2.

6.2 Electrophilic addition

For the purpose of this chapter, this will be interpreted as electrophilic addition, followed by capture of the cationic species by a nucleophile (Fig. 6.1). A simple example of this is the addition of bromine, and the mechanism and stereoelectronics of this process were covered in Chapter 2.

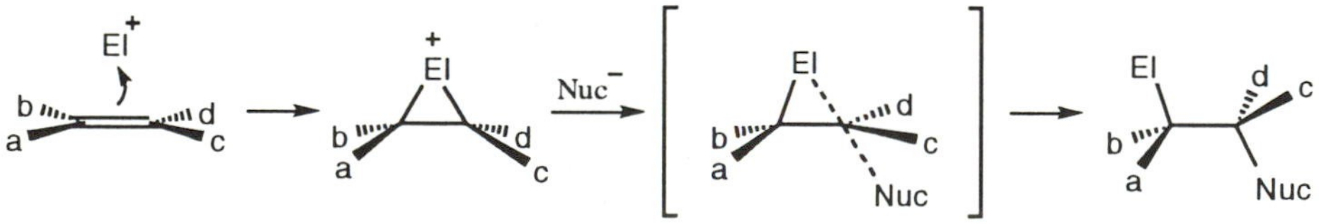

Fig. 6.1 Electrophilic addition followed by nucleophilic opening of the intermediate cation

An important general type of electrophilic addition to which is useful in stereoselective organic synthesis involves cyclization (Fig. 6.2), in which the second step in Fig. 6.1 is intramolecular. There are two possible modes of cyclization, *endo* and *exo* ring closures (Fig. 6.2).

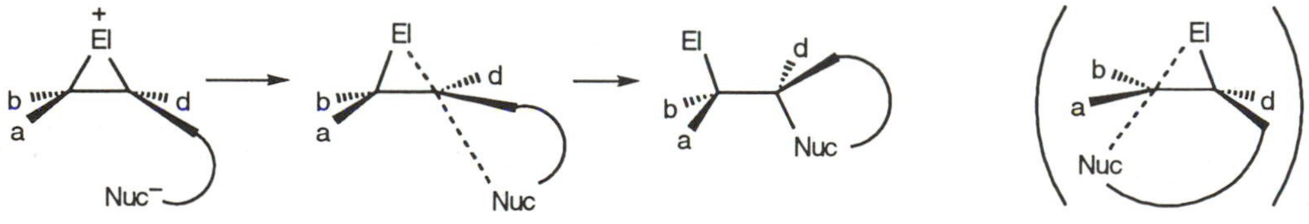

exo-cyclization

endo-Cyclization

Fig. 6.2 Electrophile induced cyclizations

It is usually much easier for the nucleophile to react via an *exo*-cyclization than by the corresponding *endo*-cyclization. This is a result of the restrictions on the nucleophile as a result of its being tethered to the double bond, and the stereoelectronic requirements of the nucleophilic attack.

endo-Cyclizations are not impossible, and structural and electronic factors can sometimes outweigh the stereoelectronic effects which favour *exo*-cyclization.

The overall stereochemistry of addition to the double bond is *trans*. For an achiral substrate undergoing this reaction, a single racemic stereoisomer should be formed, the relative stereochemistry of which will depend on the geometry of the alkene. A simple example of this process, known as iodolactonization, is shown in Fig. 6.3 and represents a valuable type of electrophile induced cyclization.

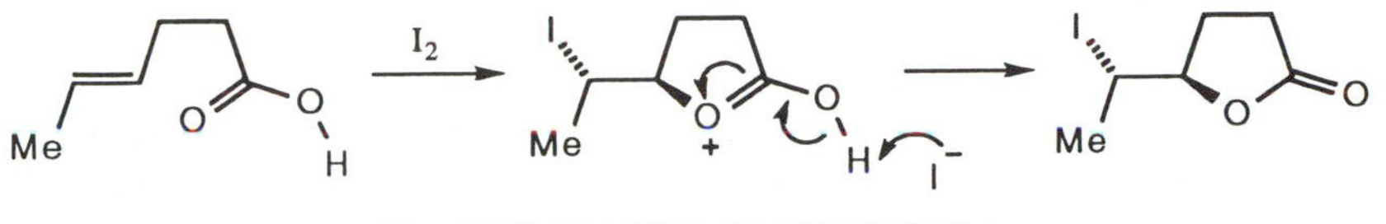

Fig. 6.3 Electrophile induced lactonization

If there is a pre-existing stereocentre in the substrate, then this can play an important part in controlling the diastereoselectivity of the reaction. Some examples of the iodolactonization of chiral substrates are shown in Fig. 6.4.

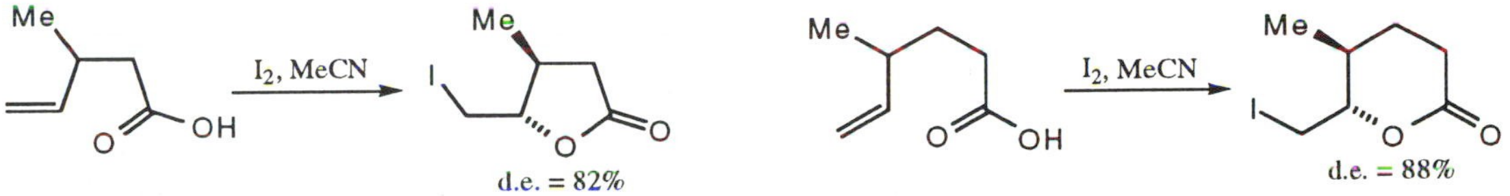

d.e. = 82%

d.e. = 88%

Fig. 6.4 Diastereoselective iodolactonization of chiral substrates

From the examples provided in Fig. 6.4, it can be seen that with the pre-existing stereocentre close to the C–C double bond, the diastereoselectivity can be very high. The reactions shown in Fig. 6.4 are probably under thermodynamic control, in each case the major diastereoisomer corresponds to the more stable product. These are examples of a highly diastereoselective reaction which is under thermodynamic control.

The reaction works with a range of nucleophiles and electrophilic reagents. Of particular interest are electrophile induced intramolecular cyclizations where the nucleophile is an alkene, which form C–C bonds (Fig. 6.5).

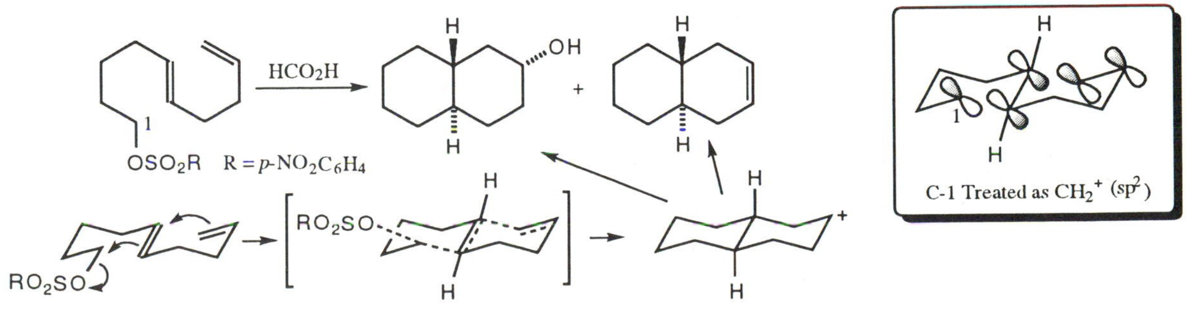

C-1 Treated as CH_2^+ (sp^2)

Fig 6.5 Correlation of *trans* alkene stereochemistry and *trans* ring fusion in polyene cyclizations

This type of reaction is of special significance, as it is analogous to that used in Nature for the biosynthesis of polycyclic natural products of the terpene family. These systems often have *trans* fused rings, and the Stork–Eschenmoser hypothesis proposed that this stereochemical feature was related to the *trans* geometry of the alkene precursors. The suggested mechanism required the new bonds to be formed more or less simultaneously from opposite faces of the alkene. This amounts to a *trans* addition to the double bond (Fig, 6.5). The reaction shown in Fig. 6.5 was a demonstration that the proposed biosynthetic reaction, in which the alkene geometry is reflected in the ring fusion stereochemistry, could be carried out in the laboratory.

6.3 Hydroboration

The mechanism and related stereoelectronic aspects of alkene hydroboration were discussed earlier in this book (Chapter 2). Overall, it is a *cis*-addition, in which the boron is attached at the more nucleophilic end of the double bond. For an achiral alkene and borane, the reaction is stereospecific. The stereoelectronic requirement of the reaction and the alkene geometry determine which (racemic) diastereoisomeric product is formed (Fig. 6.6).

Fig. 6.6 Hydroboration of an alkene

$A^{(1,3)}$ strain was responsible for the high stereoselectivity observed in the diastereoselective alkylation of chiral acyclic enolates (Chapter 4).

If the alkene is prochiral and acyclic, then high levels of diastereoselectivity can only usually be obtained if the pre-existing stereocentre is sufficiently close to the alkene. The preferred conformation of alkenes with a stereocentre adjacent to the double bond is often that in which allylic or $A^{(1,3)}$ strain is minimized (Fig. 6.7). This leads to the proposal that a reagent is likely to prefer to attack such an alkene from the face opposite to the larger group, assuming the absence of strong electronic or chelating effects due to substituents L and S. As might be anticipated from Fig. 6.7, this idea is useful in accounting for the stereoselectivity in other reactions of alkenes, including hydroboration as shown in Fig. 6.8.

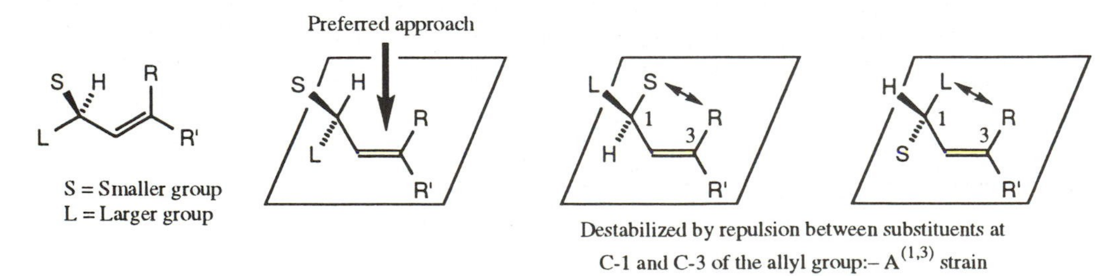

Fig. 6.7 $A^{(1,3)}$ strain considerations in the conformation of chiral acyclic alkenes

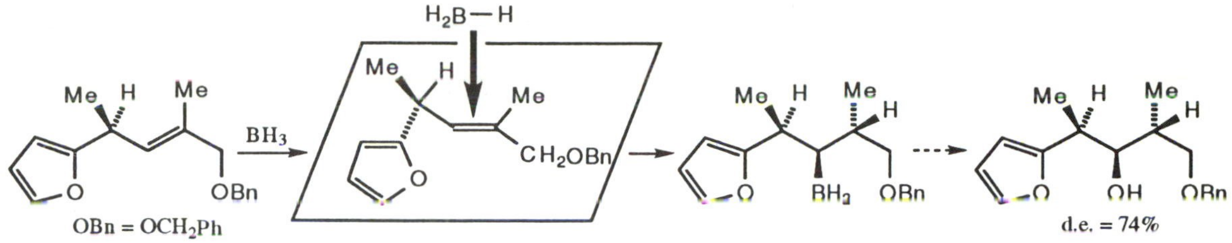

H₂B—H
Me H Me
BH₃
Me H
Me
CH₂OBn
Me H Me H
Me H Me H
O
OBn
BH₂
OH
OBn = OCH₂Ph
d.e. = 74%

Fig. 6.8 Diastereoselective hydroboration of an acyclic alkene 'controlled' by A$^{(1,3)}$ strain

Many chiral reagents have been developed for hydroboration of alkenes, a small selection of which are shown in Fig. 6.9, along with some simple examples of reactions with alkenes. The reagents derived from pinene are readily available, and easily prepared. Reagent **6.1** is more difficult to prepare, but reacts with very high enantioselectivity.

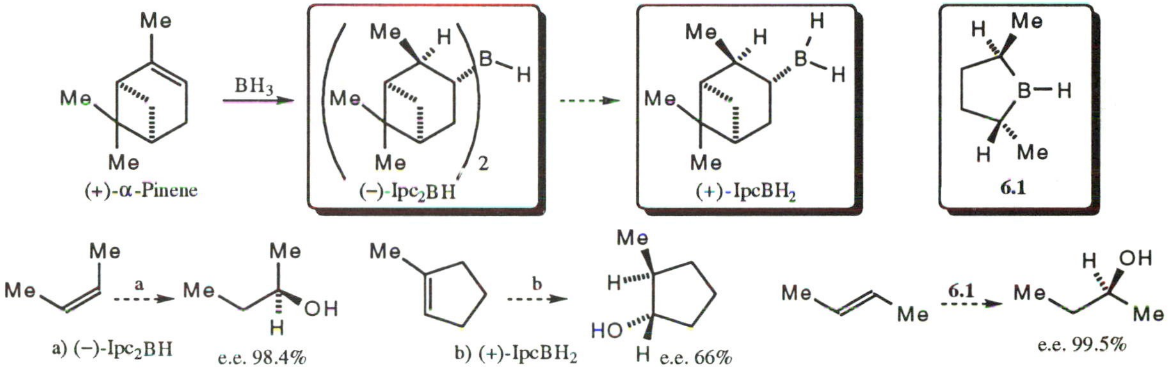

(+)-α-Pinene
BH₃
(−)-Ipc₂BH
(+)-IpcBH₂
6.1
a) (−)-Ipc₂BH
e.e. 98.4%
b) (+)-IpcBH₂
e.e. 66%
6.1
e.e. 99.5%

Fig. 6.9 Chiral reagents for hydroboration and typical examples

6.4 Conjugate additions

Nucleophilic attack on a C–C double bond usually requires that it is electron deficient, and for our purposes we will consider only α,β-unsaturated carbonyl compounds. The overall reaction is often referred to as conjugate addition (or 1,4-addition). In its most general form (Fig. 6.10), three components are involved, each of which could be chiral. The process involves two distinct reactions, addition of the nucleophile to generate an enolate, and 'trapping' of this with an electrophilic reagent.

Nucleophiles and electrophiles which are not chiral or prochiral can give rise to two possible diastereoisomeric products, although one often predominates (Fig. 6.11). The trapping step corresponds to reaction of a

Reactions of enolates with electrophiles are covered at some length in Chapter 3.

Nuc⁻
Addition
Trapping
Nuc
R
R'
El⁺
El

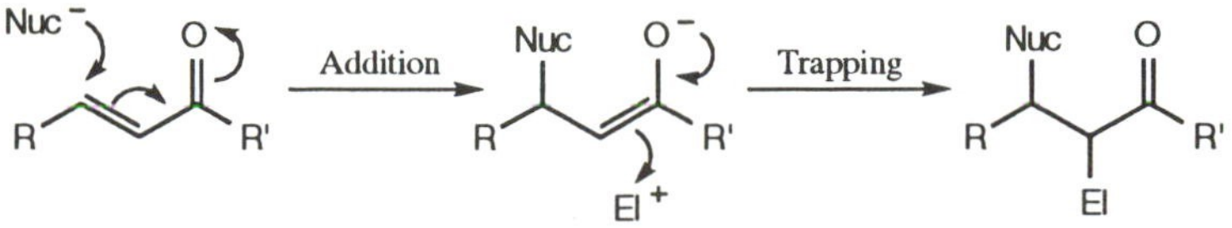

Fig. 6.10 Conjugate addition followed by enolate trapping

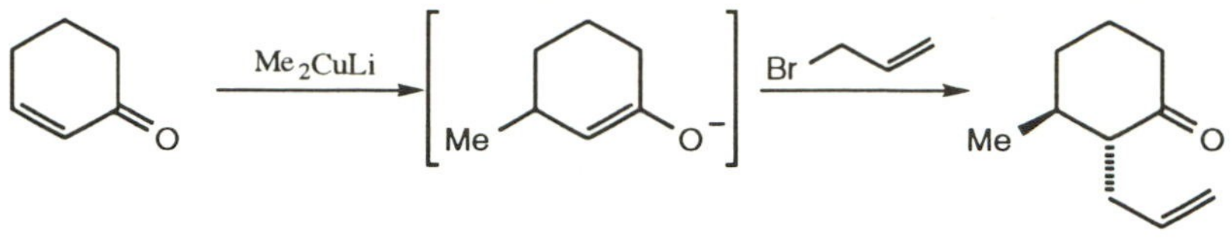

Fig. 6.11 Trapping of an enolate from the less hindered face

chiral enolate with an electrophile, which usually takes place from the less hindered face of the enolate.

If the α,β-unsaturated carbonyl compound is chiral, then addition usually takes place from the less hindered face, as shown in Fig. 6.12.

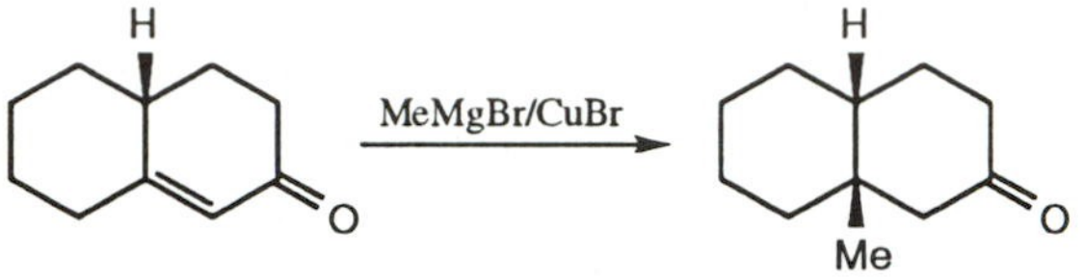

Fig. 6.12 Conjugate addition from the less hindered face

It is possible to control the stereochemistry of both the new stereocentres which are produced in a conjugate addition-trapping sequence using chiral auxiliaries. One of the most effective methods is shown in Fig. 6.13, and employs the sultam **6.2** chiral auxiliary which can also be used for enolate alkylation, aldol reaction, and Diels–Alder cycloadditions.

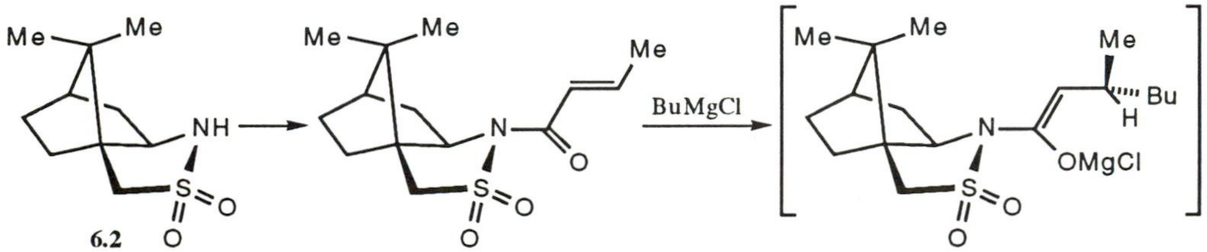

Fig. 6.13 A chiral auxiliary for conjugate addition-trapping

Enantioselective catalysts have been developed for conjugate addition reactions, but it is fair to say that there is not yet a catalyst which is capable of achieving very high enantioselectivity with a wide range of nucleophiles and α,β-unsaturated carbonyl compounds. The nickel catalyst **6.3** will catalyse the conjugate addition of dialkylzinc reagents to some acyclic α,β-unsaturated ketones with good e.e., as shown in Fig. 6.14.

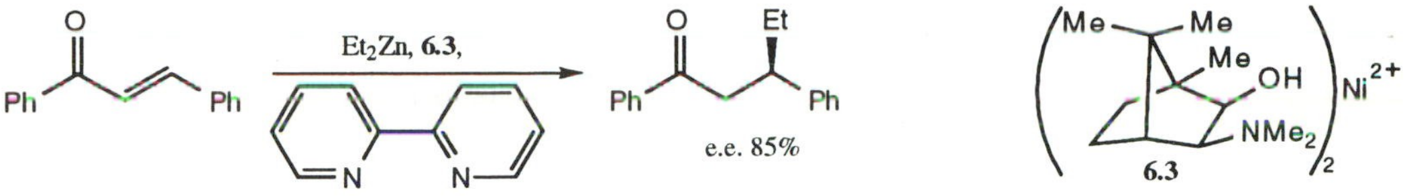

Fig. 6.14 Enantioselective catalysis of conjugate addition

6.5 Diels–Alder cycloaddition

The Diels–Alder cycloaddition is one of the most powerful general methods in organic synthesis. Two new C–C bonds, up to four new stereocentres, one double bond, and a new six membered ring are all formed in the reaction. It is a concerted thermal 4+2 cycloaddition, and some aspects of the mechanism have been discussed in Chapter 2. When both components are prochiral, two diastereoisomers can be produced, known as the *endo*- and *exo*-adducts (Fig. 6.15). Although an *endo*-approach is usually more congested, and the *endo*-adduct is thermodynamically less stable (more hindered), this is often the kinetic product.

The transition state for *endo*-cycloaddition can be stabilized by overlap of π orbitals on group B in the dienophile with the diene HOMO. Such 'secondary orbital overlap' is not possible in an *exo*-transition state.

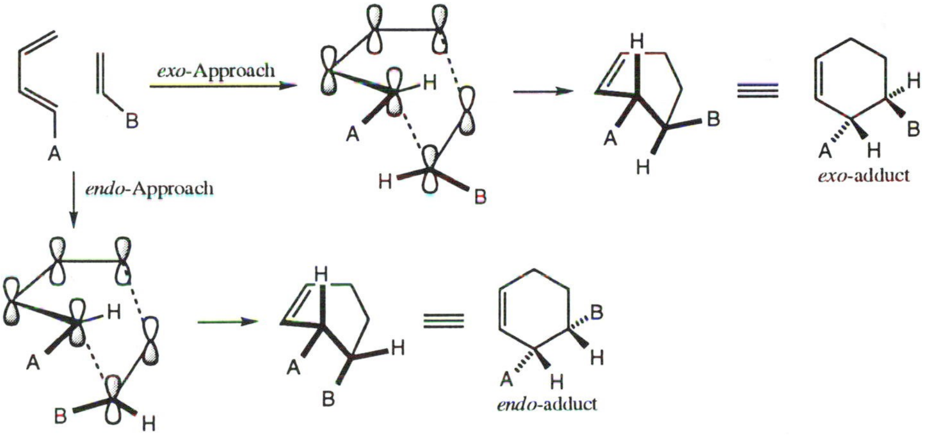

Fig. 6.15 Formation of *endo*- and *exo*-adducts in a Diels–Alder cycloaddition

It follows from above that the geometry of diene and dienophile will be 'preserved' in the adduct, and that the kinetic product will usually be the *endo*-adduct. This general scheme is illustrated in Fig. 6.16.

The regiochemistry of the cycloaddition is usually controlled by the 'relative sizes' of the p orbitals which comprise the LUMO and HOMO

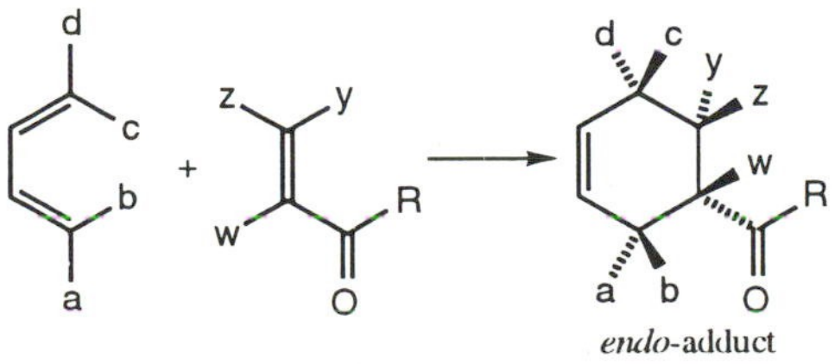

Fig. 6.16 General scheme for a Diels–Alder cycloaddition

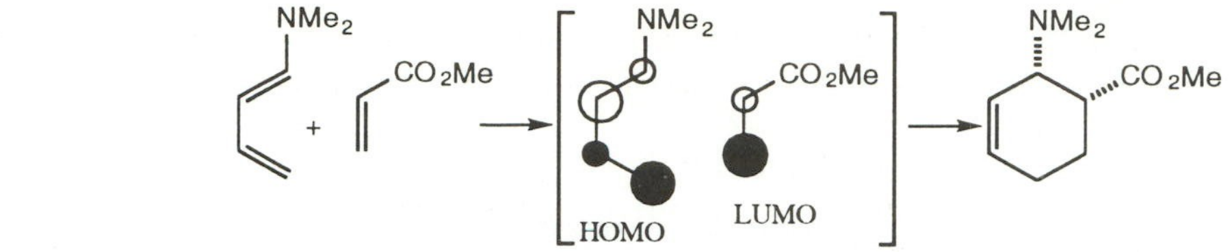

Fig. 6.17 Regioselective *endo* Diels–Alder cycloaddition

involved. More accurately, these are described as the orbital coefficients. The most common type of Diels–Alder cycloadditions involves electron deficient dienophiles. The LUMO for such a system is distorted as shown in Fig. 6.17, with the larger coefficient at the end of the alkene remote from the electron withdrawing group. A similar analysis of the HOMO of a diene possessing an electron donating group at one terminus reveals the largest coefficient to be at the end remote from the group. Best overlap in the reaction is achieved by interaction of the ends of the reactants which possess the larger coefficients. This is illustrated in Fig. 6.17.

If the two faces of the diene are diastereotopic, then the dienophile will usually react from the less hindered face, as in Fig. 6.18.

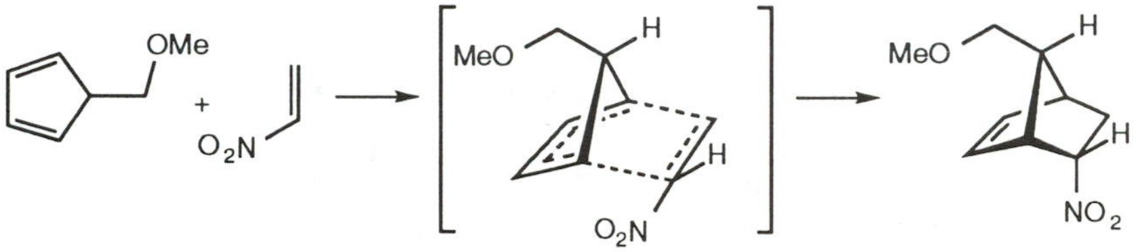

Fig. 6.18 *endo* Diels–Alder cycloaddition from less hindered face of the diene

Given the importance of the Diels–Alder cycloaddition in organic synthesis, it is not surprising that much effort has gone into developing efficient enantioselective methods. Many chiral auxiliaries are available, for example the oxazolidinones which we have encountered several times already can be very effective. In the presence of more than one mole equivalent of the Lewis acid dimethylaluminium chloride, a cationic complex is formed (Fig. 6.19. This dienophile is now extremely electron deficient, and held rigidly as the aluminium chelate. *endo*-Cycloaddition from the less hindered face then provides the major product in high d.e.

The fact that the Diels–Alder cycloaddition is mediated by or catalysed by

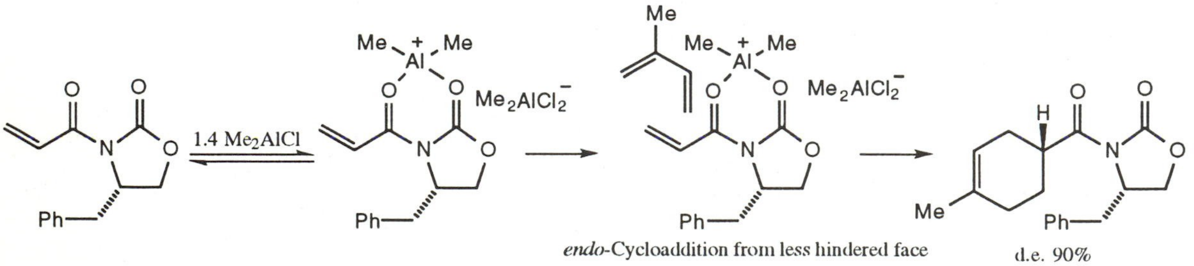

Fig. 6.19 A chiral auxiliary controlled Diels–Alder cycloaddition

Lewis acids means that the opportunity for enantioselective reaction presents itself. As with the chiral auxiliaries, there are a number of enantioselective catalysts which work well.

The chiral aluminium complex **6.4**, related to reagents which are used in carbonyl additions, enolate functionalization, aldol reactions, and some rearrangements, is an effective catalyst for the reaction of dienes and *N* aryl maleimides (Fig. 6.20).

e.e. >97%

Fig. 6.20 Asymmetric catalysis using **6.4**

6.4

A different type of enantioselective catalytic process involves a copper(II) complex of the ligand **6.5**, itself easily prepared from commercially available components. This catalyst is capable of achieving high e.e. as demonstrated by the reaction shown in Fig. 6.21.

6.5

Ar = 2,6-Dichlorophenyl

6.5 (10 %), $Cu(OSO_2CF_3)_2$ (9%)

e.e. 92%

Fig. 6.21 Asymmetric catalysis using **6.5** as a chiral ligand

6.6 Cyclopropanation

Cyclopropanation of alkenes usually involves reaction with a carbene or its equivalent. In practice the carbene is often associated with a metal, and referred to as a carbenoid. In the reaction, the geometry of the alkene is usually conserved, but this is not always the case. There are many methods for the generation of carbenoids, but we will consider only a very limited selection. A typical, example (the Simmons-Smith reagent **6.6**) is illustrated in Fig. 6.22, and reactions of this reagent are strongly directed by hydroxyl groups in the substrate (Fig. 6.22). In the absence of hydroxyl groups, as might be expected, carbenoids react with chiral alkenes from the less hindered face of the molecule.

$CH_2I_2 \xrightarrow{Zn} I{-}CH_2{-}ZnI$ (**6.6**)

d.e. >98%

Fig. 6.22 Hydroxyl directed cyclopropanation of an alkene

Decomposition of diazo compounds is catalysed by transition metals and involves metal carbenoids. Diazo esters are often employed, and the carbenoids will undergo efficient cyclopropanation reactions (Fig. 6.23).

Typically, salts or complexes of copper and rhodium are used.

As with the Simmons-Smith reagent, the alkene geometry is usually preserved, and reaction takes place from the less hindered face of the alkene where appropriate. One proposed mechanism involves formation of a four membered ring containing the metal, and collapse to the cyclopropane (Fig. 6.23).

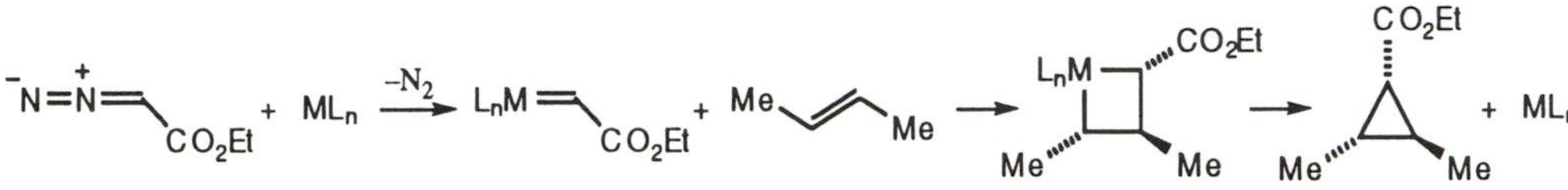

Fig. 6.23 Cyclopropanation using diazo compounds and metal salts or complexes

The sequence illustrated in Fig. 6.23 suggests that enantioselective catalysts should be possible if a metal species with appropriate chiral ligands was used. This is the case and a large number of efficient enantioselective catalysts have been developed (Fig. 6.24).

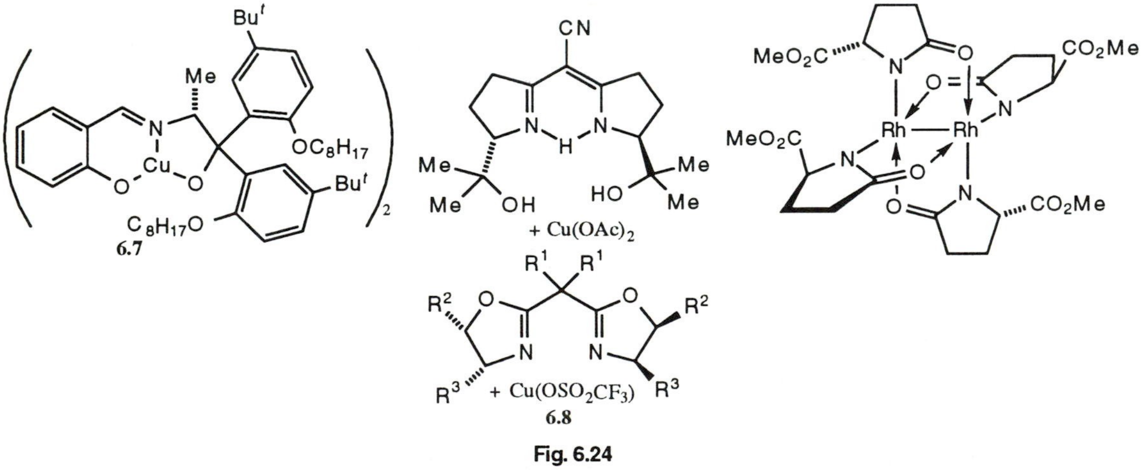

Fig. 6.24

One of the driving forces behind the development of such catalysts is the relative importance of compounds such as **6.9** (Fig. 6.25), which is a (+)-*trans*-chrysanthemic acid ester, an example of the pyrethroids, which is an important class of insecticides.

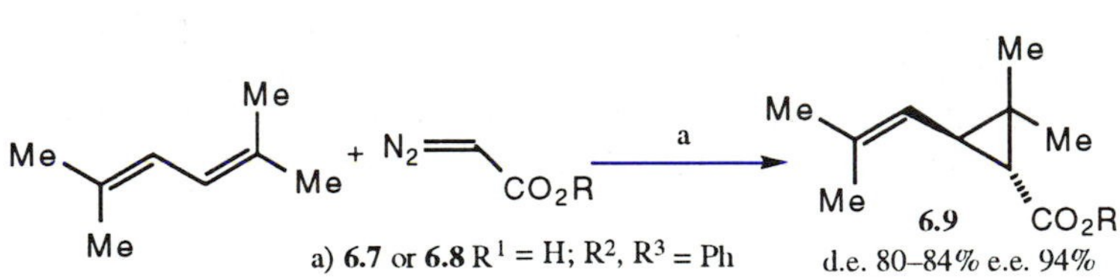

a) **6.7** or **6.8** $R^1 = H$; R^2, $R^3 = Ph$ d.e. 80–84% e.e. 94%

Fig. 6.25 Enantioselective cyclopropanation using a chiral catalyst

6.7 Problems

6.1 Assuming that the silyl group ($PhMe_2Si$) is the largest group, and that $A^{(1,3)}$ strain is an important controlling factor in the reaction, which alcohol would you expect to obtain on hydroboration/oxidation of the allylsilane **F1**?

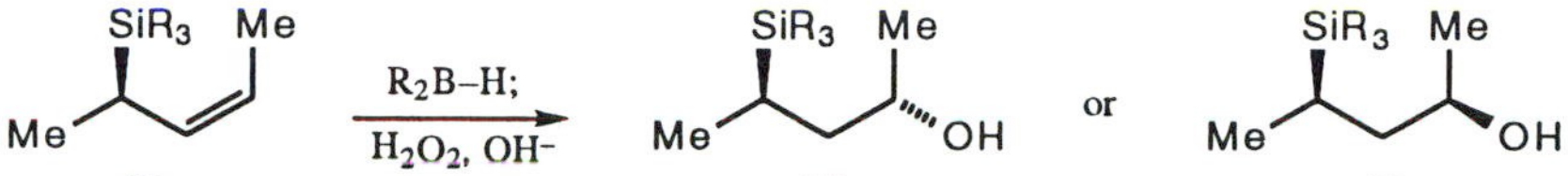

SiR_3 Me; Me; **F1**; R_2B–H; H_2O_2, OH^-; SiR_3 Me; Me; OH; **F2**; or; SiR_3 Me; Me; OH; **F3**

$SiR_3 = PhMe_2Si$

6.2 Explain the stereochemical outcome of the reaction in which **F4** undergoes conjugate addition to give **F5**.

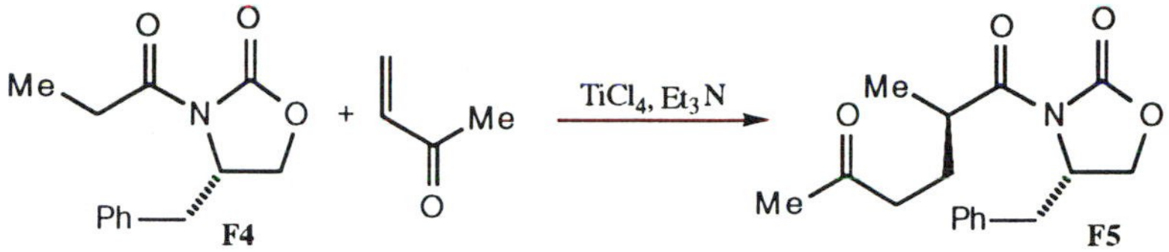

O O; Me; N; O; Ph; **F4**; +; Me; O; $TiCl_4$, Et_3N; O O; Me; O; N; O; Me; Ph; **F5**

6.3 Using the model shown in Fig. 6.19, predict the stereochemistry of the intramolecular Diels–Alder reaction of **F6**.

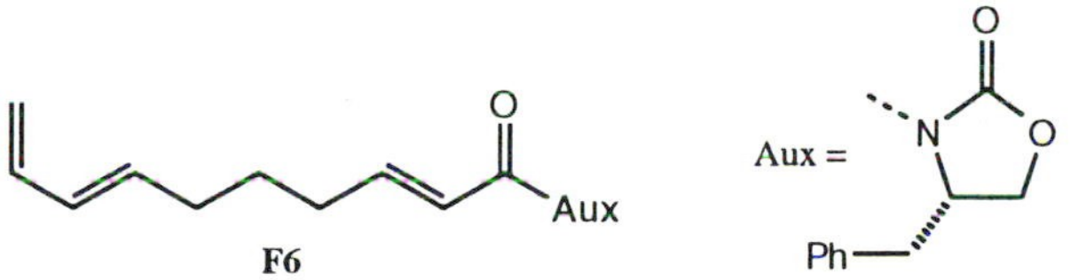

O; Aux; **F6**; Aux =; O; N; O; Ph

Suggestions for further reading

For cyclization reactions involving stereoselective additions to C–C double bonds see Chapters 5 and 6 of *Asymmetric Synthesis*, (ed. J. D. Morrison), Vol. 3, Academic Press, New York, 1984. For a discussion of $A^{(1,3)}$-strain as a controlling factor in stereoselective reactions see R. W. Hoffmann, *Chem. Rev.*, 1989, **89**, 1841–1860. J. Leonard, *Contemporary Organic Synthesis*, 1994, **1** 387–416, presents a detailed review of conjugate additions. The Diels–Alder reaction is covered in detail in many standard text books on organic chemistry. For a detailed analysis, see Chapter 4 of I. Fleming, *Frontier Orbitals and Organic Chemical Reactions*, Wiley, New York, 1977. A detailed review of 'directed' organic reactions is provided by D. A. Evans, *Chem. Rev.*, 1993, **93**, 1307–1370.

7 Reduction

7.1 General considerations

The two most commonly encountered types of reductions in stereoselective organic synthesis are the reduction of alkenes, and the reduction of carbonyl groups. Usually these are very different types of reaction (although occasionally the same reagent will do both), and each encompasses a very large area of chemistry. We will consider some of the more important aspects of these reactions from the point of view of stereoselective synthesis, along with a few examples of their application in organic synthesis.

7.2 Reduction of C–C double bonds

The most widely used method of alkene reduction is catalytic hydrogenation, and the reaction is applicable to a very wide range of types of alkenes. In general terms, the alkene is exposed to an atmosphere of hydrogen gas (or an alternative source of hydrogen) in the presence of a metal catalyst, often palladium, platinum, nickel, rhodium, or ruthenium. In heterogeneous hydrogenation, the catalyst is either the finely dispersed metal itself, or the metal adsorbed onto a support. Alternatively, soluble metal complexes can be used, which leads to homogeneous hydrogenation.

Although heterogeneous hydrogenation has been known for much longer than the homogeneous version, the mechanistic details are still relatively unclear. Hydrogen molecules are adsorbed onto the metal, probably forming metal-hydrogen σ bonds (Step 1, Fig. 7.1). It is also known that the alkene is adsorbed onto the metal surface, presumably through interaction of its π and π^* orbitals with appropriate orbitals on the metal surface (Step 2, Fig. 7.1). A hydrogen atom is then transferred to the adsorbed alkene, to form an intermediate with a metal-carbon σ bond (Step 3, Fig. 7.1). This can react with a further hydrogen atom to give the alkane (Step 4, Fig. 7.1), which then desorbs from the catalyst surface.

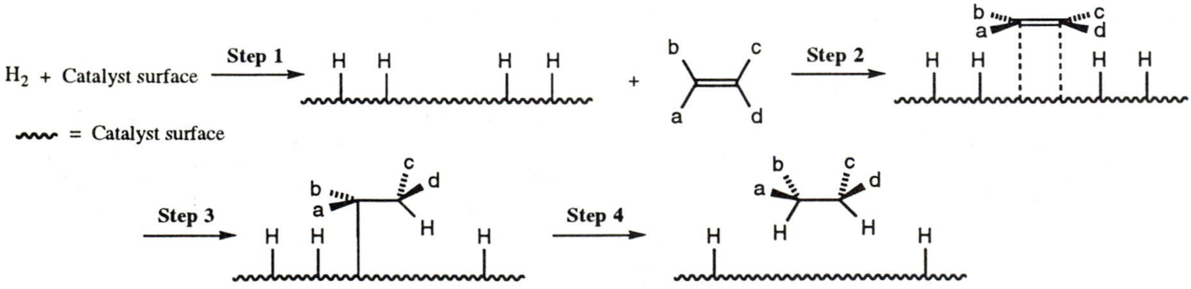

H_2 + Catalyst surface —Step 1→ ... Step 2 → ... Step 3 → ... Step 4 →

(wavy line) = Catalyst surface

Fig. 7.1 Some key steps in heterogeneous hydrogenation

In most cases, both hydrogen atoms are transferred to the same face of the double bond, resulting in a *syn* addition. As the initial 'π' bonding to the surface usually takes place from the less hindered face of the alkene (where applicable), the overall process amounts to *syn* addition of hydrogen to the less hindered face of the alkene.

Heterogeneous hydrogenation of alkenes is often stereoselective, giving the product of *syn* addition, as in the reduction of **7.1** (Fig. 7.2). However, the situation can be rather more complicated, as in some cases substantial amounts of the product of *anti* addition can be formed, as is the case in the reduction of **7.2**.

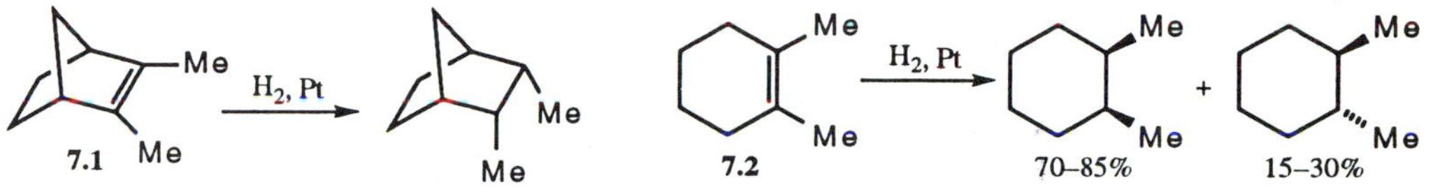

7.1 → (H_2, Pt); 7.2 → (H_2, Pt) 70–85% + 15–30%

Fig. 7.2 Heterogeneous hydrogenation of alkenes

Heterogeneous hydrogenation can be directed by hydroxyl groups in the substrate, and this effect can dominate the stereochemical outcome. Reduction of the alcohol **7.3** gives mainly the product of hydrogen addition from the same face as the hydroxyl group (Fig. 7.3). If the CH_2OH group is changed to an ester, then reduction takes place from the opposite face. The hydroxyl group binds to the surface of the catalyst, so presenting the 'upper' face of the alkene preferentially, but the ester group is not capable of such binding.

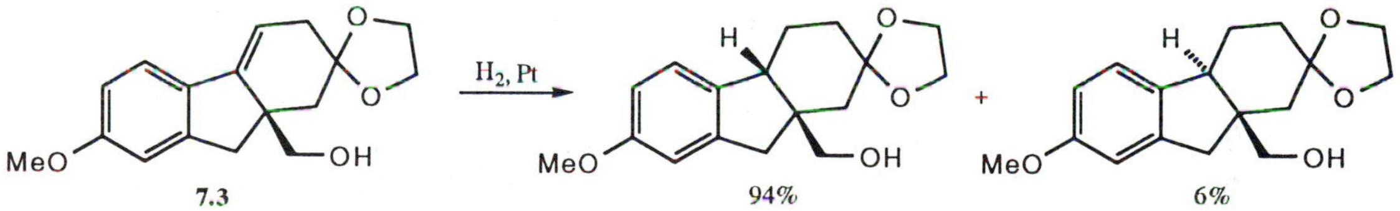

7.3 → (H_2, Pt) 94% + 6%

Fig. 7.3 Hydroxyl directed heterogeneous hydrogenation of an alkene

This type of hydroxyl direction is particularly evident in homogeneous hydrogenations. The metal is used in the form of a soluble complex, often involving phosphine ligands. The overall process is similar to that outlined for heterogeneous hydrogenation, in that it is thought to take place by reaction of H_2 with the metal to give M–H bonds, complexation of the alkene to the metal (by π bonding), followed by transfer of hydrogen from the metal to the alkene. One of the most commonly used catalysts is

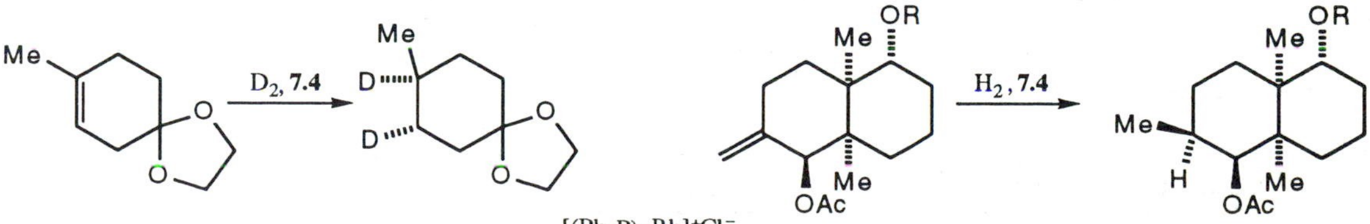

(D_2, **7.4**); (H_2, **7.4**)

$[(Ph_3P)_3Rh]^+Cl^-$

7.4

Fig. 7.4 Homogeneous hydrogenation with Wilkinson's catalyst

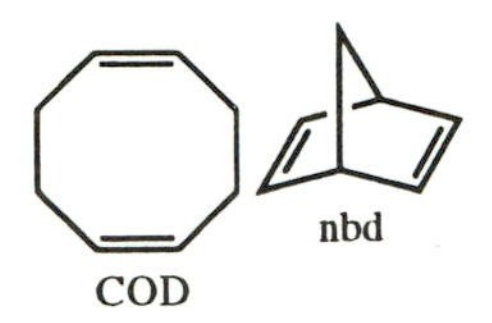

Wilkinson's catalyst **7.4** (Fig. 7.4), a cationic rhodium complex which usually results in *syn* addition of hydrogen from the less hindered face of the alkene (Fig. 7.4).

Hydroxyl directing effects are particularly evident when cationic complexes of iridium and rhodium are used. The complexes which we will consider include dienes as ligands initially, which are removed by hydrogenation under the reaction conditions, and play no part in the reaction. These dienes, 1,4-cyclooctadiene and norbornadiene are usually abbreviated to COD and nbd respectively. The structures of two such catalysts are shown in Fig. 7.5.

$[(Ph_3P)Ir(COD)py]^+ PF_6^-$

$[Rh(nbd)(diphos\text{-}4)]^+ BF_4^-$

Fig. 7.5 Structures of iridium and rhodium catalysts

The examples of directed homogeneous hydrogenation shown in Fig. 7.6 illustrate the effectiveness of this method The high diastereoselectivity observed in reduction of the methyl ether **7.5** suggests that hydrogen bonding is unlikely to be responsible for the directing effect. The metal is cationic, and co-ordination of the oxygen is much more likely to take place through one of its lone pairs.

a) H_2, $[(Ph_3P)Ir(COD)py]^+ BF_4^-$

7.5

Fig. 7.6 Directed homogeneous hydrogenation of cyclic alkenes

Highly diastereoselective directed hydrogenation of acyclic alkenes such as **7.6** is also possible, and the observed diastereoselectivity has been rationalized on the basis of a complex in which the metal is bound to the alkene and the oxygen, in which allylic strain is minimized (**7.7**, Fig. 7.7).

The mechanism proposed for this type of directed homogeneous hydrogenation is illustrated in Fig. 7.8. This overall reaction scheme embodies many of the important elements of homogeneous hydrogenation,

7.6 H_2, $[Rh(nbd)(diphos\text{-}4)]^+ BF_4^-$ **7.7** d.e. 98%

Fig. 7.7 Directed homogeneous hydrogenation of acyclic alkenes

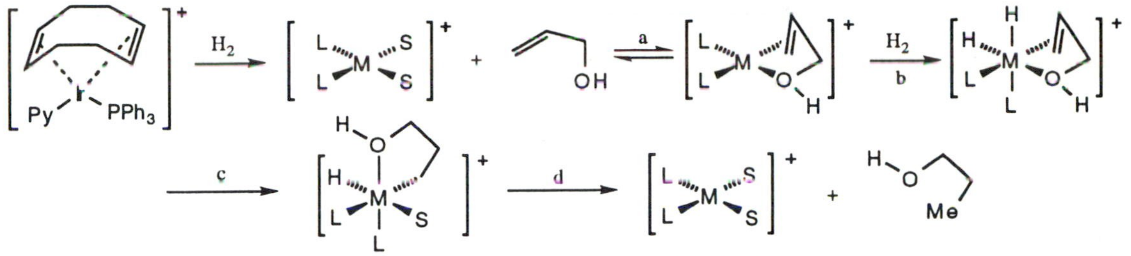

S = Solvent molecule, L = Ligand (Py & PPh_3)

a) Coordination of the alkene b) Oxidative addition of H_2 to metal c) Insertion of M–H bond into C=C
d) Reductive elimination (loss of M–H and formation of C–H bond)

Fig. 7.8 Mechanistic scheme for directed homogeneous hydrogenation

directed or not. It could be modified to accommodate reduction of alkenes which do not contain an oxygen simply by replacing the M–O bond with a M–S (metal–solvent) bond.

The mechanistic scheme shown in Fig. 7.8 is typical of homogeneous hydrogenation, and as can be seen, the ligands remain attached to the metal throughout the reaction. This means that if the alkene is prochiral, and the ligands are chiral, we have the possibility for an enantioselective catalytic procedure. This can be achieved in practice, and is a very powerful method for the asymmetric synthesis of a range of organic compounds.

A wide range of chiral phosphine ligands has been used in such enantioselective hydrogenation, but we will consider just two examples, DIPAMP (**7.8**) and BINAP (**7.9**) (Fig. 7.9). Both enantiomers are available, so that either enantiomeric hydrogenation product can be obtained.

Cationic rhodium complexes of DIPAMP are particularly useful for the asymmetric synthesis of amino acids and their derivatives from 'dehydro-α-acylamino acids' (such as **7.10**, Fig. 7.9). This approach finds use in a commercial synthesis of L-DOPA.

So far, we have covered some of the important aspects and applications of catalytic reduction of alkenes, and the remaining part of this chapter is concerned with the other important type of reduction used in organic synthesis, reduction of carbonyl groups.

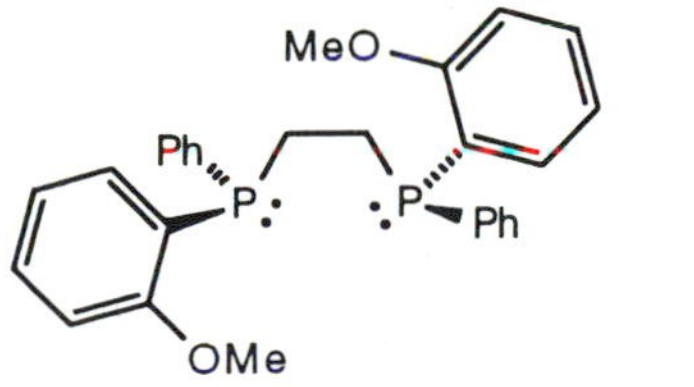

7.8 (*R*,*R*)-DIPAMP (Chiral at phosphorus)

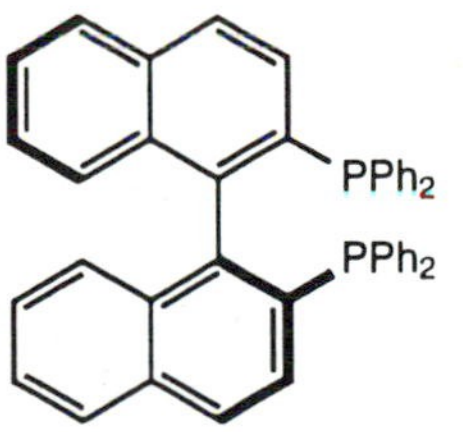

7.9 (*S*)-BINAP

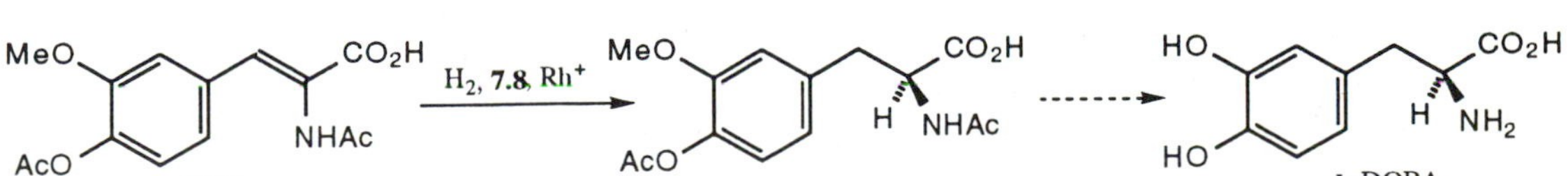

7.10 — H_2, **7.8**, Rh^+ → ⇢ L-DOPA

Fig. 7.9 Catalytic asymmetric synthesis of L-DOPA

7.3 Reduction of carbonyl groups

The reductions which we will consider involve the addition of hydride, using complex hydrides, to ketones. These reactions can be viewed as nucleophilic addition of hydride, and there is much similarity between these reactions and the nucleophilic additions discussed in Chapter 3.

Variants of $NaBH_4$ and $LiAlH_4$:– $Zn(BH_4)_2$; $LiBH_4$; $Li[B(Bu^i)_3H]$; $Na[B(OAc)_3H]$; $Li[Al(Bu^t)_3H]$

The reduction of chiral ketones is usually carried out using either $NaBH_4$ of $LiAlH_4$, although there is quite a range of variants on these simple reagents. The reactivity and the bulk of the reducing agents of the various hydride reagents vary greatly, and can affect the diastereoselectivity.

As with other nucleophilic additions, steric effects tend to dominate, and in the absence of strong steric effects, electrostatic and dipole interactions can be important. The formalism which is used for acyclic chiral ketones is the Felkin–Ahn model (Chapter 3, Section 3.2), and is outlined in Fig. 7.10.

Felkin–Ahn model for reduction of chiral acyclic ketones

d.e. 60%

d.e. 48%

Fig. 7.10 Diastereoselective reduction of chiral acyclic ketones

The reduction of cyclohexanones again follows the addition of nucleophiles (Chapter 3, Fig. 3.6) and tends to take place by axial attack, as most complex hydrides are relatively small nucleophiles. Axial substituents in the 3 position with respect to the carbonyl group disfavour axial attack, and can result in the product of equatorial reduction (Fig. 7.11).

When the structure of the ketone allows for chelation of the metal cation of the hydride reagent, then very high levels of diastereoselectivity can be obtained. If zinc borohydride is used (Fig. 7.12), the ketone acts as a bidentate chelating ligand for the zinc, and the hydride then reacts from the less hindered face of the carbonyl group.

d.e. up to 64%

d.e. up to 60%

Fig. 7.11 Reduction of cyclohexanones

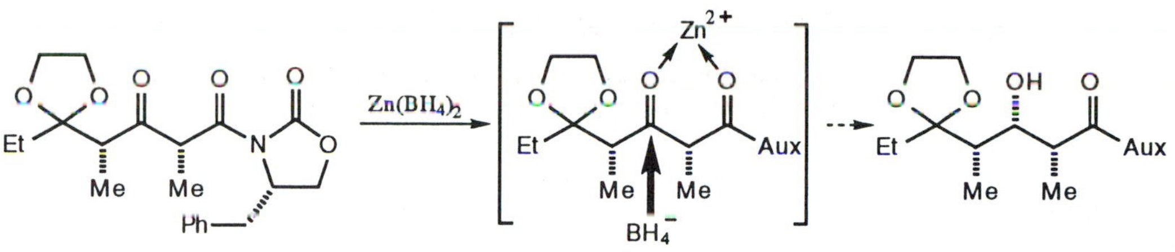

Fig. 7.12 'Chelation controlled' reduction of a ketone

The final type of reduction which we will cover is that of prochiral ketones to enantiomerically enriched secondary alcohols. Many chiral reagents have been developed and used, based on complex hydrides or boranes, and selected examples are provided in Fig. 7.13. In these reactions, the chiral species is in effect 'attached' to a 'standard' reducing agent, and is required in at least stoichiometric quantities. Although it is often possible to achieve high enantiomeric excesses in some cases, these methods have largely been superseded by methods using enantioselective catalysis.

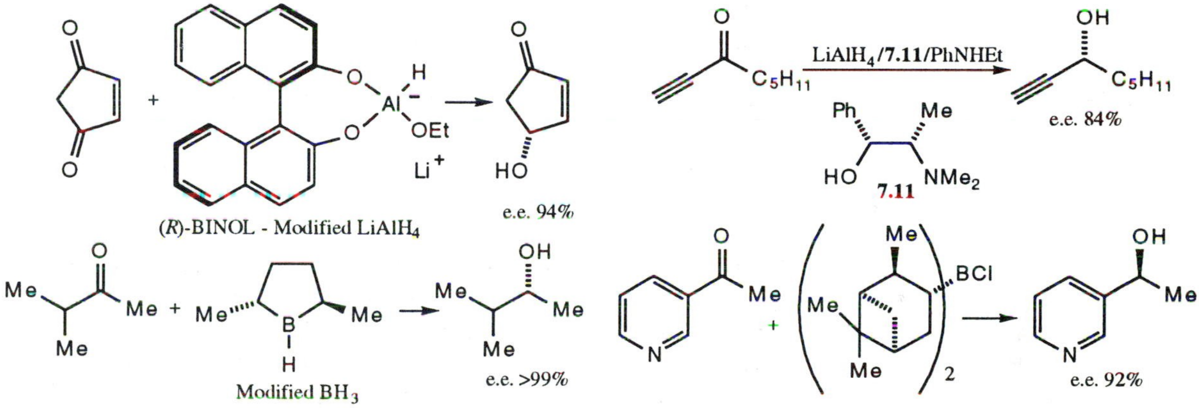

Fig. 7.13 Enantioselective ketone reduction using chiral reagents

The enantioselective reduction of a wide range of prochiral ketones can be achieved using the enantiomeric catalysts **7.12** and **7.13** (Fig. 7.14) and borane as the stoichiometric reducing agent. These are known as CBS reductions. The catalysts are easy to make, and are commercially available, making CBS reduction an attractive method for asymmetric synthesis.

The term 'CBS' is derived from the initials of the scientists who developed the catalysts; Corey, Bakshi, and Shibata.

The catalyst brings the ketone and the borane together (**7.14**, Fig. 7.15), with the borane and the ketone bound on the less hindered face, and with the larger of the substituents of the ketone directed away from the complex.

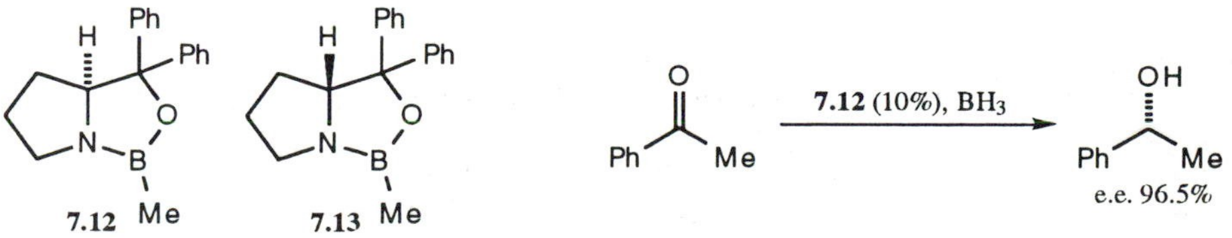

Fig. 7.14 Catalysts and conditions for CBS reduction of prochiral ketones

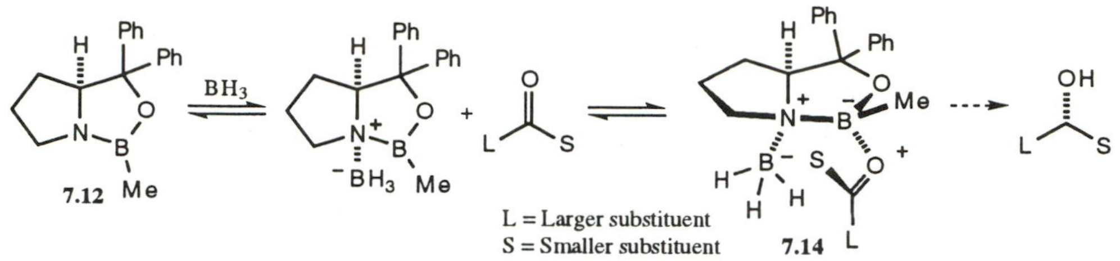

L = Larger substituent
S = Smaller substituent

Fig. 7.15 Proposed mode of action of CBS catalyst

In general, CBS reduction of ketones gives high enantioselectivity for both acyclic and cyclic ketones when there is a reasonable difference in size between the two substituents. Selected examples are illustrated in Fig. 7.16, and as can be seen, ketones of various types react successfully, and other functional groups can be tolerated.

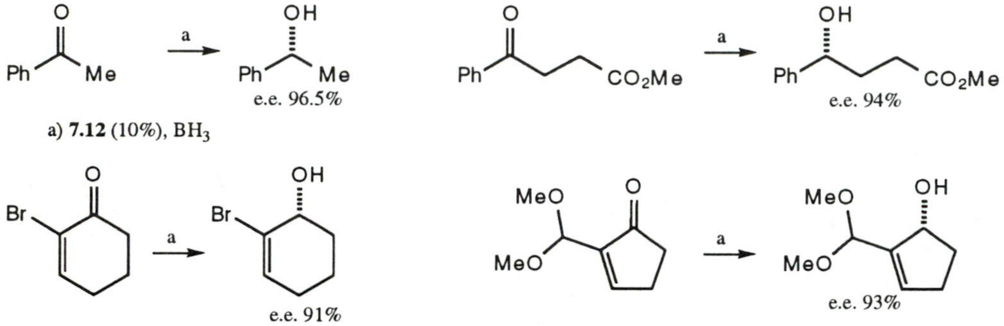

a) 7.12 (10%), BH_3

e.e. 96.5%; e.e. 94%; e.e. 91%; e.e. 93%

Fig. 7.16 CBS reduction of ketones

If the structure of the ketone is such that it is capable of acting as a bidentate chelating ligand, then enantioselective reduction is possible by hydrogenation in the presence of ruthenium(II) complexes of BINAP (see Fig. 7.9). The oxygen of the ketone acts as one donor ligand, and the other group completes the chelation. As can be seen from the examples in Fig. 7.17 this 'other' group can be either an amine, a hydroxyl, or a carbonyl group, and the chelate ring can be either five or six membered.

As a carbonyl group can act as the 'other' ligand for chelation of the ketone, it is not surprising that diketones can be excellent substrates for this type of enantioselective reduction. The product of reduction of one carbonyl

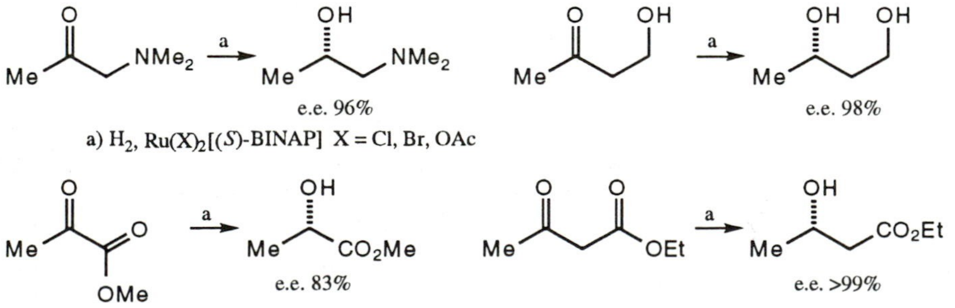

a) H_2, $Ru(X)_2[(S)\text{-BINAP}]$ X = Cl, Br, OAc

e.e. 96%; e.e. 98%; e.e. 83%; e.e. >99%

Fig. 7.17 Enantioselective reduction of ketones with Ru(II)-BINAP complexes

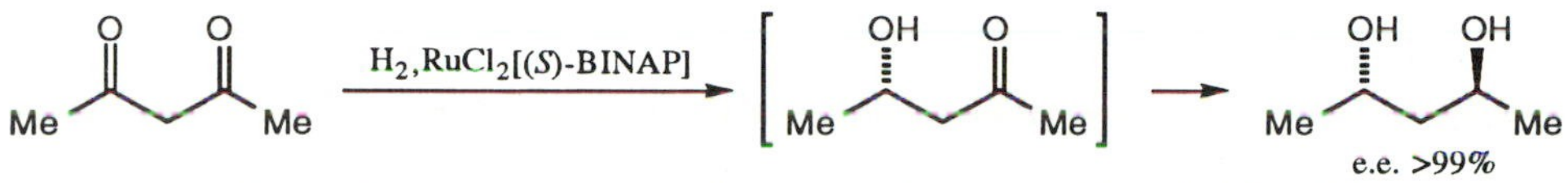

Fig. 7.18 Enantioselective reduction of a 1,3-diketone with $RuCl_2$[[(*S*]-BINAP]

group in a diketone, a hydroxyl group, is itself capable of chelation, and so the reaction proceeds to give the diol. This process is both highly enantioselective and diastereoselective (Fig. 7.18).

The catalysts which we have discussed for the enantioselective reduction of ketones act by binding both the substrate and the reagent in some way, and this is similar to the way in which enzymes operate. As carbonyl reduction is a common reaction in Nature, it is not surprising that there are many enzymes which can be used for the enantioselective reduction of ketones. In practice, one of the easiest ways of achieving this is to use the enzymes present in actively fermenting yeast. In general, steric effects dominate, and a simple model accounts for the observed enantioselectivity in most reductions of this type (Fig. 7.19).

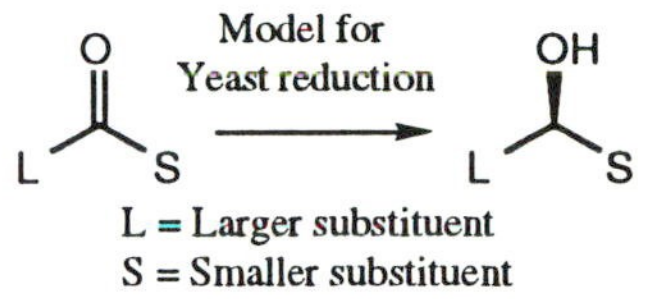

Model for Yeast reduction

L = Larger substituent
S = Smaller substituent

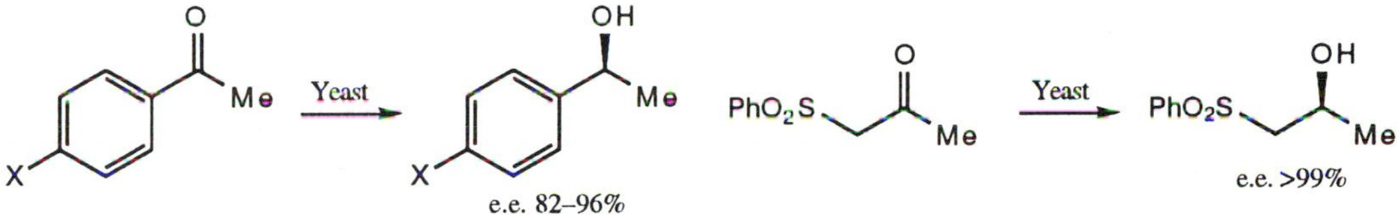

Fig. 7.19 Yeast reduction of prochiral ketones

β-Keto esters are also excellent substrates for yeast reductions (Fig. 7.20). A β-keto ester which possesses a substituent between the two carbonyls is chiral, and the two enantiomers are in rapid equilibrium via the enol. If such a β-keto ester is subjected to yeast reduction, then often one of the enantiomers is reduced much more rapidly than the other. As this takes place, the keto–enol equilibrium is continuously re-established, so the slow reacting enantiomer is converted into the fast reacting enantiomer, which of course is reduced. In this way, we can obtain greater than 50 per cent yield of essentially enantiomerically pure product starting from a racemate (Fig. 7.20).

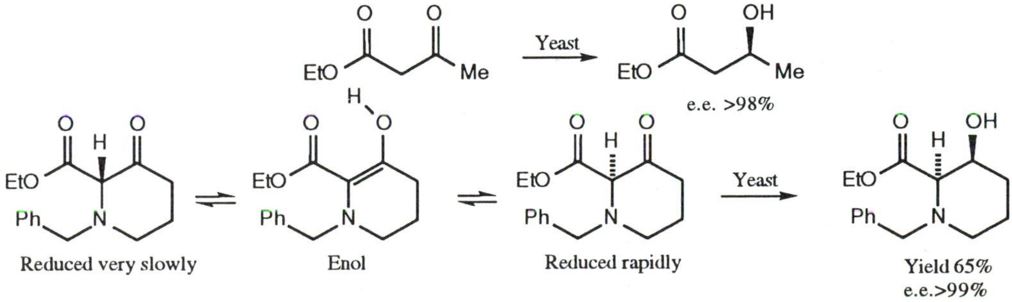

Fig. 7.20 Yeast reduction of racemic β-keto esters

7.4 Problems

7.1 Hydrogenation of **G1** gives (*R*)-citronellol. What would be the product if **G2** was used in place of **G1**? What product would you expect if (*R*)-BINAP was used in place of (*S*)-BINAP in the reduction of **G1**?

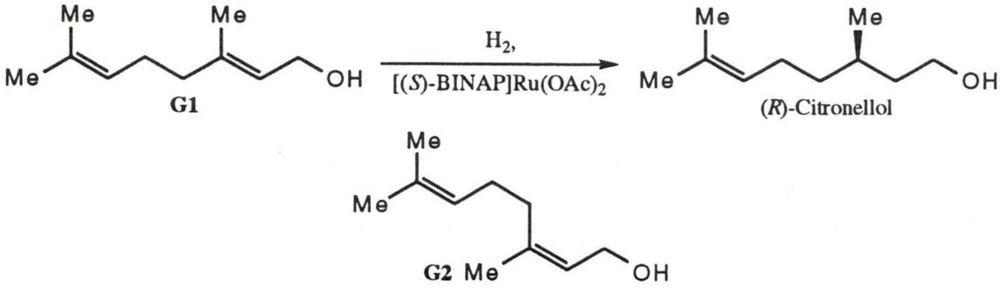

7.2 Account for the selective formation of the reduction products **G3** and **G4**.

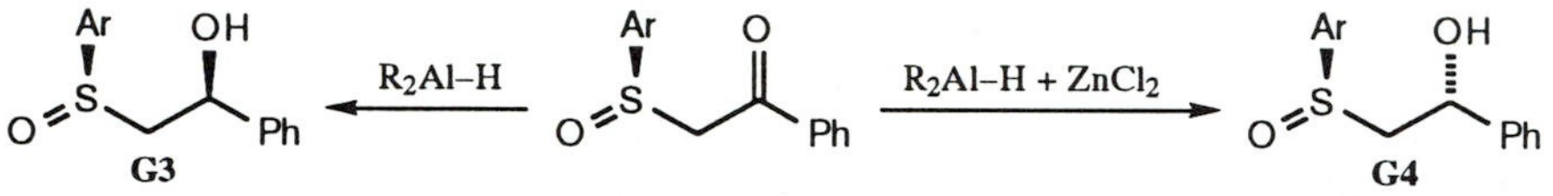

7.3 Why does the reduction of **G5** give **G6** selectively?

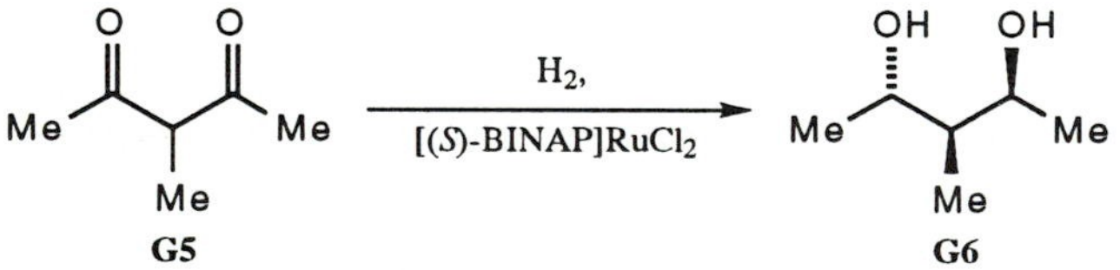

Suggestions for further reading

Catalytic hydrogenation is covered in Chapters 2 and 3 of *Asymmetric Synthesis*, (ed. J. D. Morrison), Vol. 5, Academic Press, New York, 1985, and the basics can be found in Chapter 5 of F. A. Carey and R. J. Sundberg, *Advanced Organic Chemistry*, 3rd edition, Part B, Plenum, 1990. Examples of 'directed' hydrogenation can be found in D. A. Evans, *Chem. Rev.*, 1993, **93**, 1307–1370. Detailed analysis of hydride additions to ketones and aldehydes can be found in Chapters 2–5 of *Asymmetric Synthesis*, (ed. J. D. Morrison), Vol. 2, Academic Press, New York, 1983, and V. K. Singh, *Synthesis*, 1991, 605–617, provides a review of the enantioselective reduction of ketones.

8 Oxidation

8.1 General considerations

The oxidation of alkenes can be a powerful method for stereoselective synthesis, and the reactions which we will consider in this chapter are oxidations of alkenes to give epoxides, 1,2-diols, or 1,2-amino alcohols.

8.2 Epoxidation

Epoxides are usually obtained by oxidation of an alkene, and are particularly useful as intermediates for organic synthesis, as they can undergo highly selective ring openings with nucleophilic reagents. The epoxidation process usually involves retention of configuration of the alkene, and the ring opening reaction is usually an S_N2 process. This makes epoxidation followed by nucleophilic ring opening a sequence whose stereochemical outcome is highly predictable, as might be expected for two sequential stereospecific reactions. The overall process is outlined below (Fig. 8.1).

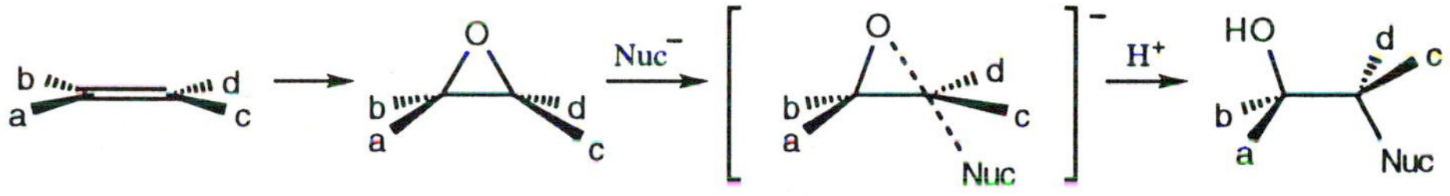

Fig. 8.1 Stereochemistry of epoxidation followed by ring opening

For the epoxidation of simple prochiral double bonds (i.e. those without strongly electron withdrawing substituents) to give racemic epoxides, the normal reagent is an organic peroxyacid, often meta-chloroperoxybenzoic acid (*m*CPBA, **8.1**, Fig. 8.2). The mechanism of this reaction has been discussed briefly in Chapter 2 (Fig. 2.18). As neither the reagent nor substrate is chiral, and no chiral catalyst is used, the epoxide **8.2** must be racemic.

If the alkene is chiral, then a diastereoselective reaction is possible. In general, the epoxidation will take place preferentially on the less hindered face of the alkene. If there is a hydroxyl group adjacent to the alkene, then this can 'direct' the peroxyacid by hydrogen bonding. This was the first type of reaction to be recognized as being subject to such 'direction'. A typical example of this type of reaction is shown in Fig. 8.3.

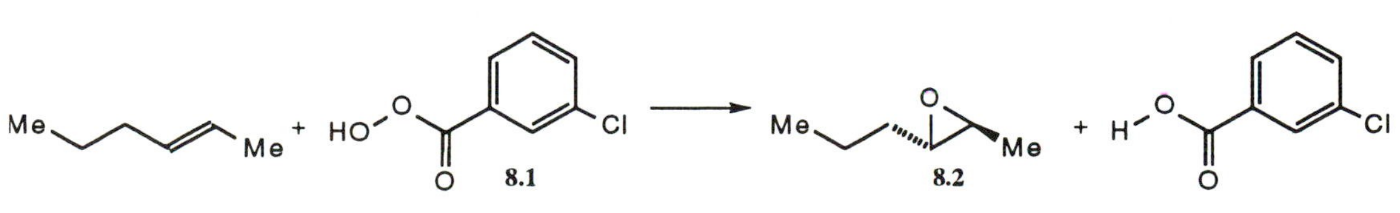

Fig. 8.2 Epoxidation of a simple alkene with a peroxyacid

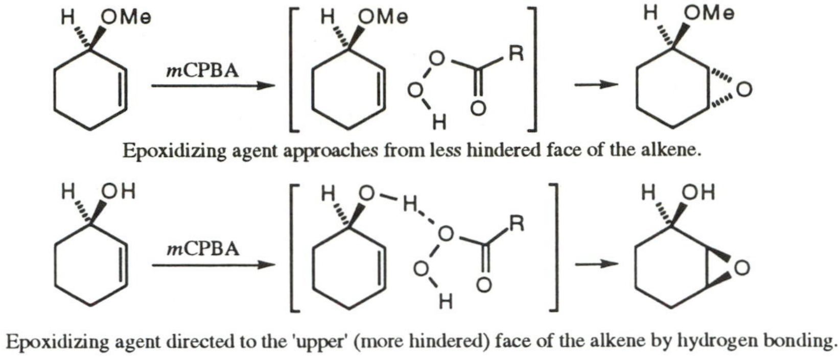

Epoxidizing agent approaches from less hindered face of the alkene.

Epoxidizing agent directed to the 'upper' (more hindered) face of the alkene by hydrogen bonding.

Fig. 8.3 Epoxidation of a prochiral alkene with *m*CPBA

When the epoxide is attached to a six membered ring which itself has a rigid conformation, then reaction with a nucleophile is both stereospecific, and highly regioselective. The epoxide opens to give the product of *trans-diaxial* opening, in which the incoming nucleophile and the hydroxyl group are both axial (Fig. 8.4).

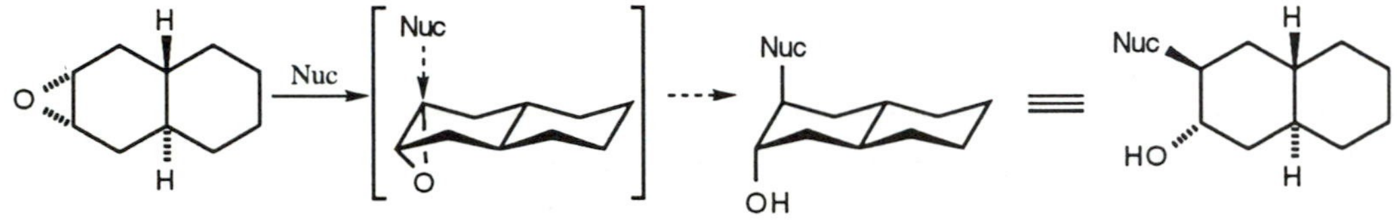

Fig. 8.4 Ring opening of an epoxide on a rigid six membered ring

Highly diastereoselective epoxidation of chiral acyclic alkenes is also possible, particularly if there is a hydroxyl group adjacent to the alkene. In this type of epoxidation (Fig. 8.5), the reacting conformation of the alkene is important, as is 'direction' by hydrogen bonding.

When racemic allylic alcohol **8.3** reacts with *m*CPBA the two diastereoisomeric epoxides **8.4** and **8.5** are formed with **8.4** being predominant (Fig. 8.5). This can be understood by considering representations of the two reacting conformations **8.6** and **8.7** which lead to the diastereoisomeric epoxides. In both, the peroxyacid is directed by hydrogen bonding to the allylic hydroxyl group, but **8.6** is likely to be lower energy than **8.7**, as in the latter there will be considerable $A^{(1,3)}$ strain due to the close proximity of the two methyl groups

See Chapters 4 and 6 for other examples of the role of $A^{(1,3)}$ strain in stereoselective reactions.

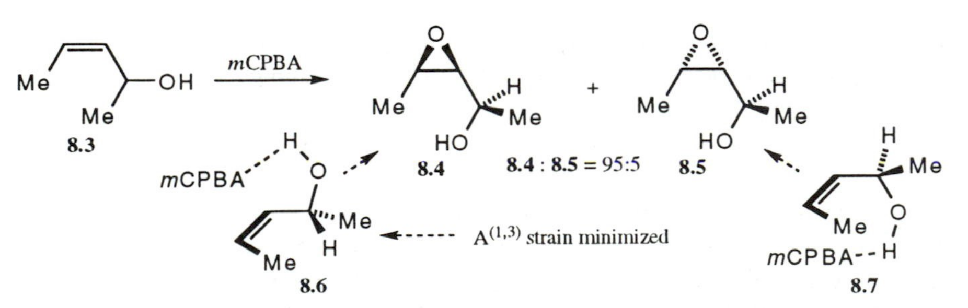

Fig. 8.5 Diastereoselective epoxidation of an acyclic, chiral, allylic alcohol

Epoxidation is not confined to the use of *m*CPBA. In the presence of certain transition metal complexes as catalysts (Mo, V, W, or Ti), alkyl hydroperoxides such as *tert*-butyl hydroperoxide (TBHP, **8.8**, Fig. 8.6) will epoxidize alkenes. This approach is useful for allylic alcohols, as the hydroxyl group can 'direct' the epoxidation by co-ordination to the catalyst. An added advantage of these catalysts is that the hydroxyl group can exert its directing effect even when it is rather remote from the alkene.

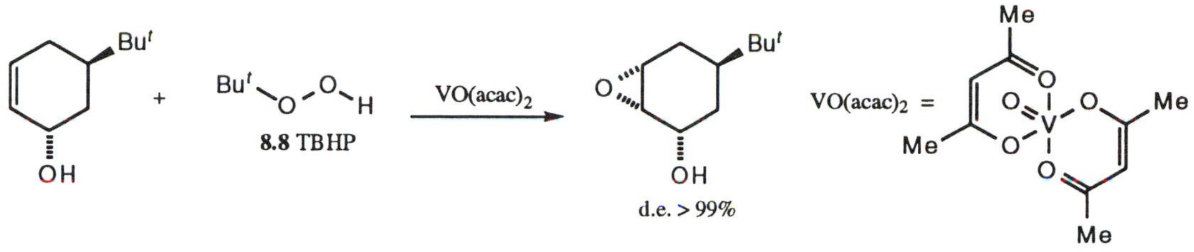

8.8 TBHP

VO(acac)$_2$

d.e. > 99%

VO(acac)$_2$ =

Fig. 8.6 Epoxidations with *tert*-butyl hydroperoxide (TBHP)

So far, all the discussion of epoxidation reactions has concerned either prochiral or racemic substrates, so the product epoxides must be racemic. Great advances have been made in the development of methods for producing epoxides in high e.e., and this is an area of organic synthesis where enantioselective catalysis has made a major impact.

A method for the enantioselective catalytic epoxidation of prochiral allylic alcohols, known as Sharpless Asymmetric Epoxidation (Fig. 8.7), uses catalytic amounts of an ester of tartaric acid such as diisopropyl tartrate **8.9** (DIPT), or diethyl tartrate **8.10** (DET), and titanium tetraisopropoxide (Ti(OPr^i)$_4$), with TBHP (**8.8**) as the stoichiometric oxidizing agent.

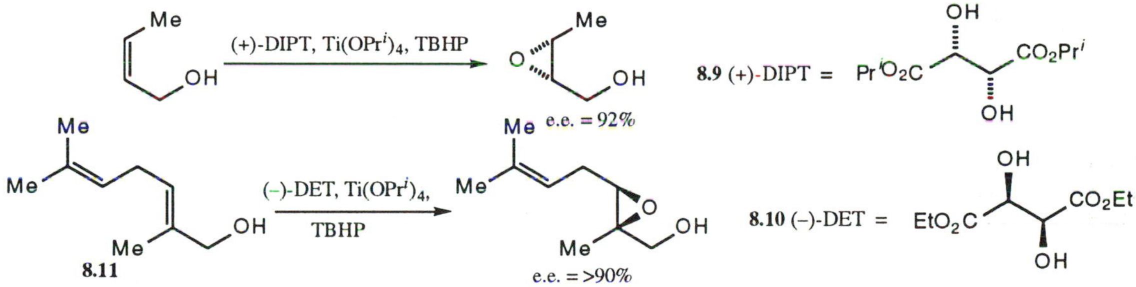

(+)-DIPT, Ti(OPr^i)$_4$, TBHP

e.e. = 92%

8.9 (+)-DIPT =

(–)-DET, Ti(OPr^i)$_4$, TBHP

8.11

e.e. = >90%

8.10 (–)-DET =

Fig. 8.7 Sharpless Asymmetric Epoxidation of prochiral allylic alcohols

The reaction is very general, and a very wide range of prochiral allylic alcohols can be used as substrates, making an equally wide range of epoxides available in high e.e. Moreover, it is highly selective for allylic alcohols, other double bonds in the substrate are not epoxidized, as in the epoxidation of **8.11** (Fig. 8.7).

A detailed discussion of the mechanism of this reaction is beyond the scope of this book, but a brief description follows. The catalyst is derived from the chelating tartrate ligand and Ti(OPr^i)$_4$), and epoxidation takes place when the catalyst is 'loaded' with both the epoxidizing agent (TBHP) and the allylic alcohol. The product epoxide and *tert*-butanol (from the TBHP) are then replaced by more allylic alcohol and TBHP, and the cycle continues.

The catalytic complex delivers the epoxide oxygen from one face or the other of the double bond, depending on the chirality of the tartrate ligand, in a highly predictable manner, as illustrated in Fig. 8.8.

(–)-Dialkyl Tartrate, $Ti(OPr^i)_4$, TBHP

Usually e.e. > 90%

With the alkene drawn vertically, and the OH at the bottom right:–
(–)-Tartrate diesters will deliver the epoxide oxygen from above, and
(+)-Tartrate diesters will deliver the epoxide oxygen from below

(+)-Dialkyl Tartrate, $Ti(OPr^i)_4$, TBHP

Usually e.e. > 90%

Fig. 8.8 Facial selectivity of Sharpless Asymmetric Epoxidation using tartrate diesters

The Sharpless Asymmetric Epoxidation conditions can also be applied to racemic allylic alcohols such as **8.12**. The catalyst is chiral, and only one enantiomer is present, so both enantiomers of the allylic alcohol should in principle be epoxidized from the same face, as illustrated in Fig. 8.9. However, the *rates* of epoxidation of the two enantiomers will be different (the transition states are diastereoisomeric).

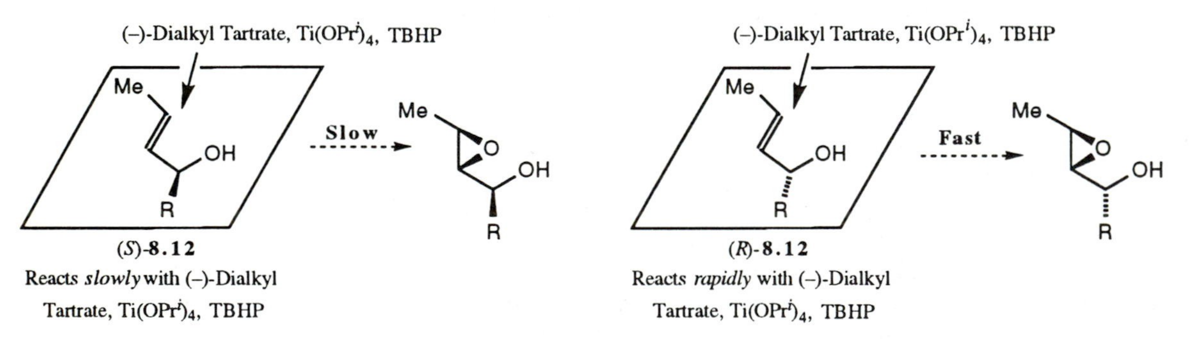

Fig. 8.9 Sharpless kinetic resolution of racemic allylic alcohols

If the rates of epoxidation of the two enantiomers are sufficiently different, and this is very often the case, then using 0.5 mole of TBHP for each mole of racemic allylic alcohol should result in essentially complete consumption of the faster reacting enantiomer (there is 0.5 mole of it in the racemate). The reaction should stop (there being no oxidant left) at the point where almost all the epoxide product derives from the faster reacting enantiomer, and almost all the unreacted allylic alcohol is the slower reacting enantiomer. This results in kinetic resolution, which is generally very efficient with this type of epoxidation. An example of this so-called Sharpless Kinetic Resolution is shown in Fig. 8.10.

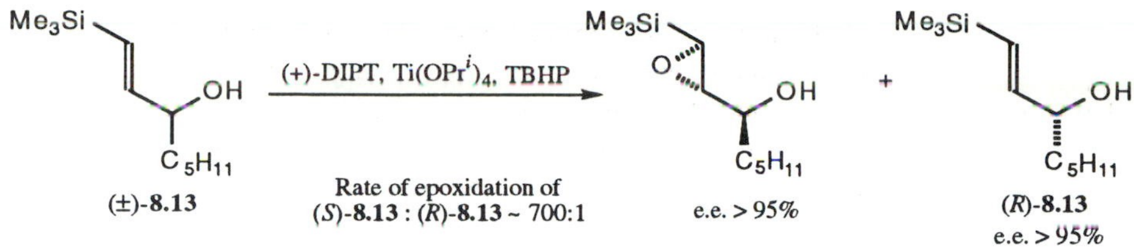

Fig. 8.10 Example of Sharpless Kinetic Resolution

Sharpless Asymmetric Epoxidation and Kinetic Resolution are particularly useful since all the reagents are commercially available and quite cheap, both enantiomers of the tartrate ligand are readily available, the reactions generally give good yields of products with high e.e., and the absolute and relative stereochemistry of the products are highly predictable. Not surprisingly, these reactions have found widespread use in organic synthesis. An example, which emphasizes the selectivity of the epoxidation step, and the regioselective and stereospecific opening of the epoxide produced, is illustrated in Fig. 8.11.

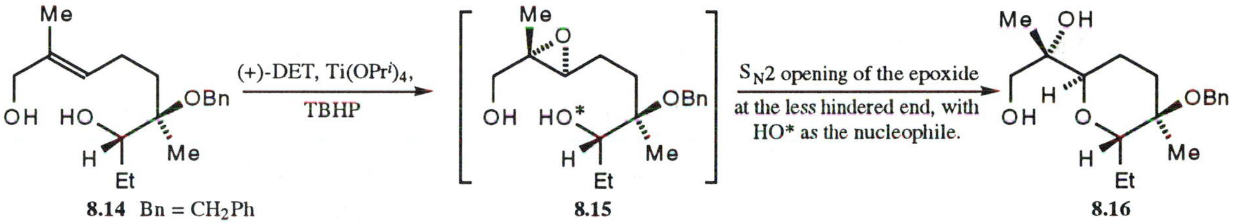

Fig. 8.11 Sharpless Asymmetric Epoxidation and *in situ* epoxide opening.

Sharpless Asymmetric Eßœpoxidation of **8.14** takes place to give the epoxide **8.15** in high e.e., and with the expected stereoselectivity. The initially produced epoxide reacts further under the reaction conditions, undergoing cyclization by an intramolecular epoxide opening with the nucleophilic hydroxyl group to give **8.16**.

Sharpless Asymmetric Epoxidation and Kinetic Resolution are extremely powerful and useful tools in organic synthesis, but they are both restricted to allylic alcohols. This selectivity can be put to good use, but is a restriction on the types of epoxide which can be prepared.

Catalytic enantioselective epoxidation of *cis*-disubstituted alkenes is possible using catalysts such as **8.17** (Fig. 8.12), and in the presence of sodium hypochlorite (bleach) under basic conditions, these will epoxidize a

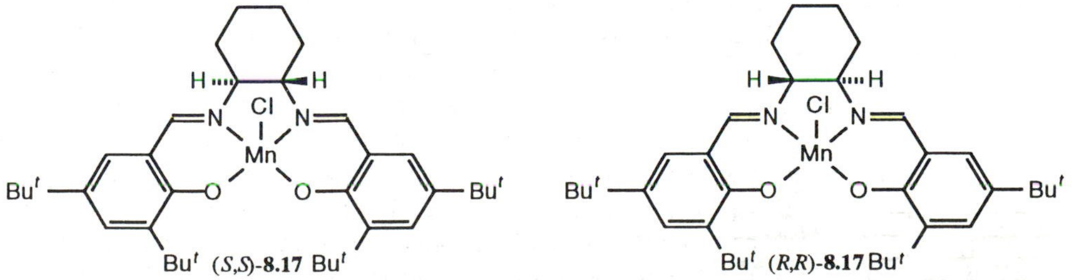

Fig. 8.12 Complexes used in the catalytic enantioselective epoxidation of *cis*-alkenes

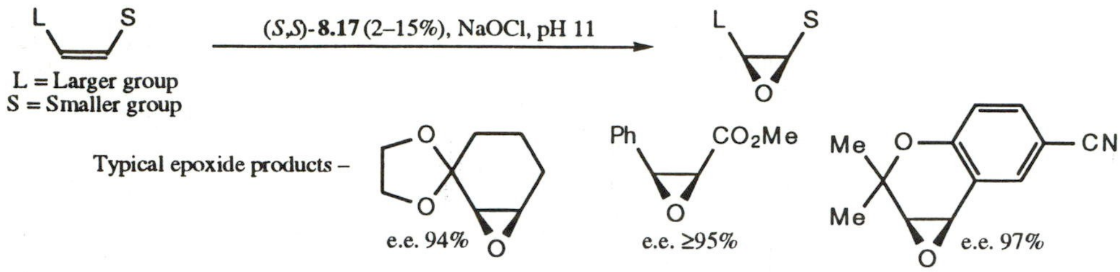

L = Larger group
S = Smaller group

(*S*,*S*)-**8.17** (2–15%), NaOCl, pH 11

Typical epoxide products –

e.e. 94% e.e. ≥95% e.e. 97%

Fig. 8.13 Enantioselective epoxidation of *cis*-alkenes

range of *cis*-alkenes. The facial selectivity of these catalysts can be predicted using the simple steric model shown in Fig. 8.13, along with some examples of the epoxides which can be prepared.

It is now possible to prepare a wide range of epoxides using enantioselective catalysis, and although all types of alkenes cannot yet be used as substrate, the range is sufficient for many synthetic purposes. A complementary type of alkene oxidation, dihydroxylation, is often highly diastereoselective, and is also possible using enantioselective catalysis. This, and aminohydroxylation of alkenes, constitute the rest of this chapter.

8.3 Dihydroxylation

Dihydroxylation of an alkene using osmium tetroxide **8.18** is a stereospecific reaction in which produces a *syn*-diol such as **8.19** (Fig. 8.14). Osmium tetroxide is usually used in catalytic quantities, along with a stoichiometric oxidant to recycle the reduced osmium species represented as **8.20** following reaction with the alkene. The reaction proceeds through a cyclic osmate ester **8.21**, which is hydrolysed under the reaction conditions, to the diol and reduced osmium species. Oxidation of the latter allows the reaction cycle to continue (Fig. 8.14). One advantage of dihydroxylation using osmium tetroxide is that it is a very general reaction, electron rich, electron deficient, and electronically 'neutral' alkenes will all undergo reaction, as will heavily substituted alkenes.

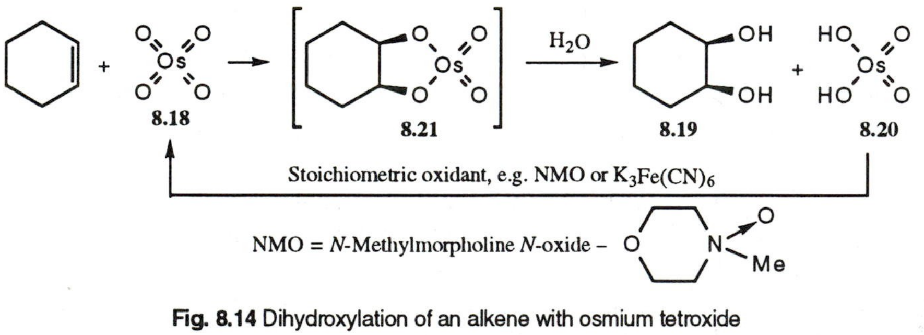

H_2O

8.18 **8.21** **8.19** **8.20**

Stoichiometric oxidant, e.g. NMO or $K_3Fe(CN)_6$

NMO = *N*-Methylmorpholine *N*-oxide –

Fig. 8.14 Dihydroxylation of an alkene with osmium tetroxide

Dihydroxylation of a chiral alkene is often reasonably diastereoselective. Steric effects can be important, with the *syn*-diol being introduced from the less hindered face of the chiral alkene (Fig. 8.15), but the stereoselectivity varies greatly depending on the structure of the alkene.

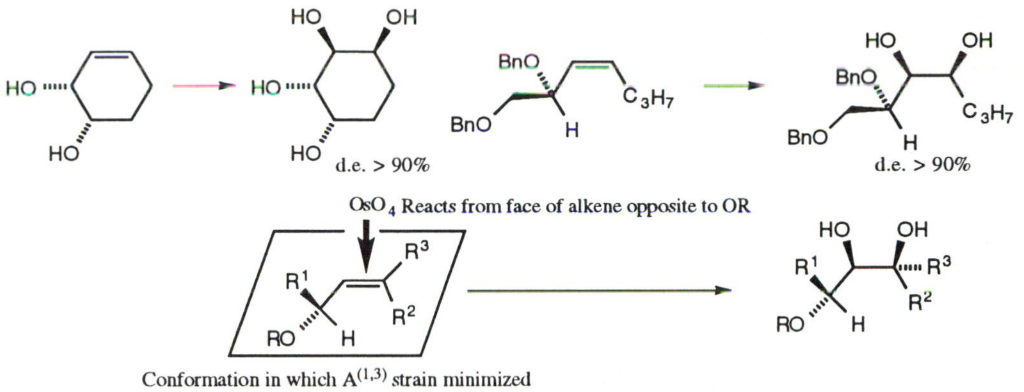

d.e. > 90%

d.e. > 90%

OsO_4 Reacts from face of alkene opposite to OR

Conformation in which $A^{(1,3)}$ strain minimized

Fig. 8.15 Diastereoselective dihydroxylation of chiral allylic alcohols and ethers

The dihydroxylation of chiral allylic alcohols and ethers often takes place with good diastereoselectivity, but unlike epoxidation, the neighbouring hydroxyl group does not direct by hydrogen bonding. One model proposed to account for the observed diastereoselectivity is shown in Fig. 8.15.

Enantioselective dihydroxylation of prochiral alkenes is possible using catalytic conditions (again developed by Sharpless) which are very efficient, and under which many different types of alkene react with very high enantioselectivity. A variety of enantiomerically pure ligands have been developed, the most widely used, **8.22** and **8.23**, are shown in Fig. 8.16. These ligands co-ordinate to the osmium, which can be osmium tetroxide or $K_2[OsO_2(OH)_4]$, with potassium ferricyanide as the stoichiometric oxidant, and potassium carbonate to control the pH. In many cases the efficiency of the process is increased by the use of methanesulfonamide.

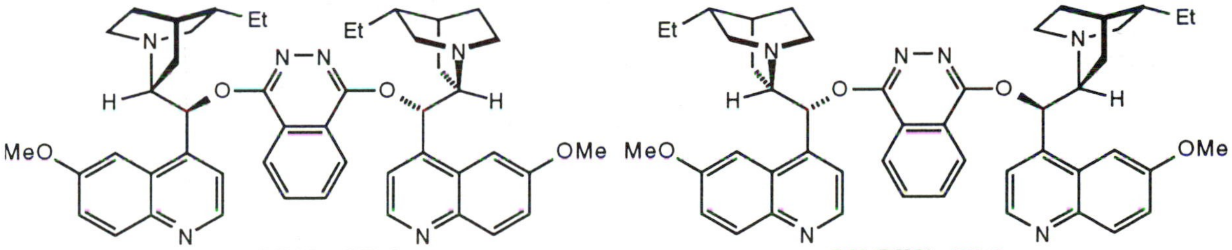

8.22 $(DHQD)_2$-PHAL

8.23 $(DHQ)_2$-PHAL

Fig. 8.16 Ligands used for Sharpless Asymmetric Dihydroxylation

Although the ligands may appear to be very complicated, in fact they are derived from either quinine or quinidine, which are readily available natural products, and the ligands are commercially available. Even more conveniently, mixtures of either ligand, $K_2[OsO_2(OH)_4]$, potassium ferricyanide, and potassium carbonate are available, making this type of Sharpless Asymmetric Dihydroxylation very simple to carry out in practice. Many types of alkene will undergo Sharpless Asymmetric Dihydroxylation, and a few examples are shown in Fig. 8.17.

The high levels of enantioselectivity, and the lack of interference from other functional groups in the alkene, combined with the ease of carrying out

a) **8.22** $(DHQD)_2$-PHAL, OsO_4 or $K[OsO_2(OH)_4]$, K_2CO_3, $K_3Fe(CN)_6$
b) **8.22** $(DHQD)_2$-PHAL, OsO_4 or $K[OsO_2(OH)_4]$, K_2CO_3, $K_3Fe(CN)_6$, $MeSO_2NH_2$

Fig. 8.17 Selected examples of Sharpless Asymmetric Dihydroxylation of alkenes

the reaction, make Sharpless Asymmetric Dihydroxylation a very useful reaction for natural product synthesis. A simple example is the synthesis of (+)-*exo*-brevicomin (Fig. 8.18), an insect pheromone.

$(DHQD)_2$-PHAL, K_2CO_3, $K_3Fe(CN)_6$, $MeSO_2NH_2$, OsO_4 → e.e. 95% → TsOH → **8.24** (+)-*exo*-Brevicomin

Fig. 8.18 Synthesis of *exo*-brevicomin using Sharpless Asymmetric Dihydroxylation

A full mechanistic understanding of the Sharpless Asymmetric Dihydroxylation process is not yet available, indeed at the time of writing there is considerable debate over some aspects of the reaction. Nevertheless, a simple pictorial 'working model' for the catalytic site, which at least accounts for the experimentally observed selectivity has been used widely, and is illustrated below (Fig. 8.19).

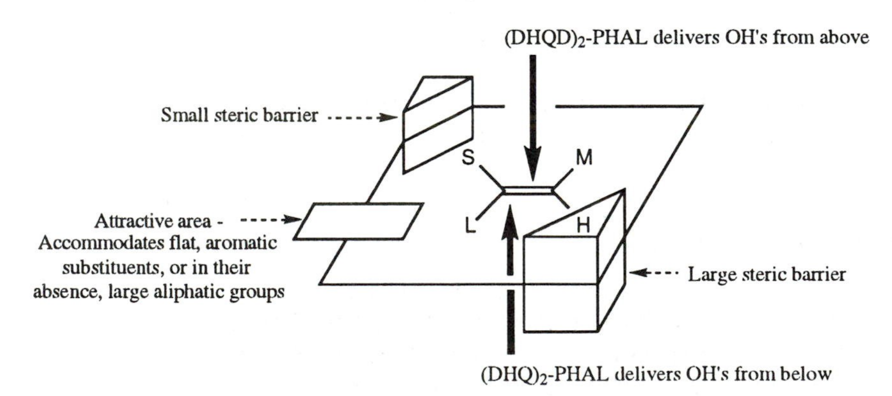

Fig. 8.19 Pictorial model for catalytic site in Sharpless Asymmetric Dihydroxylation

8.4 Aminohydroxylation

Aminohydroxylation results in the conversion of an alkene into a 1,2-aminoalcohol derivative, and can be achieved with highly enantioselectivity using the same catalytic system as that used for Sharpless Asymmetric

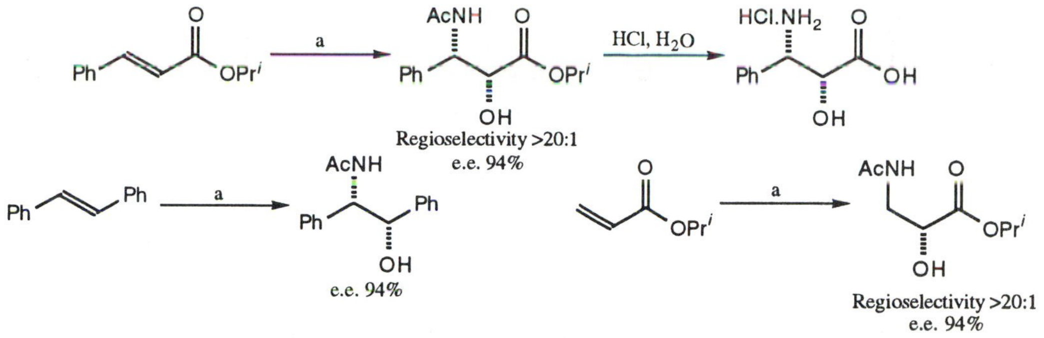

a) AcNHBr (1.1 mole), LiOH (1.07 mole), $K_2[OsO_2(OH)_4]$ (0.015 mole), $(DHQ)_2$-PHAL (0.01 mole)

Fig. 8.20 Sharpless Asymmetric Aminohydroxylation of alkenes

Dihydroxylation. A suitable nitrogen donor is used in addition to the ligand, the osmium salt, and base, and in the examples shown in Fig. 8.20, this nitrogen donor is *N*-bromoacetamide.

When the alkene is unsymmetrically substituted, there are two possible regioisomers, but as can be seen from Fig. 8.20, Sharpless Asymmetric Aminohydroxylation can be highly regioselective as well as enantioselective.

8.5 Problems

8.1 Epoxides **H1** and **H2** were prepared by Sharpless Asymmetric Epoxidation. Which enantiomer of the tartrate ester was used in each case?

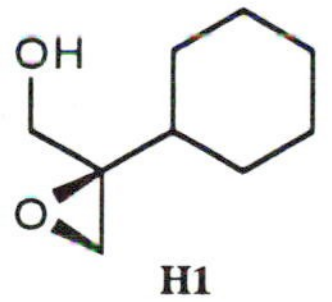

H1 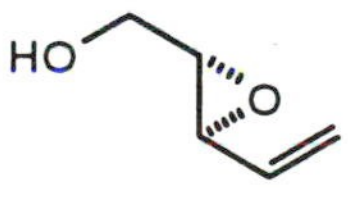 **H2**

8.2 Would either enantioselective epoxidation or enantioselective dihydroxylation be particularly useful in the stereoselective oxidation of **H3**?

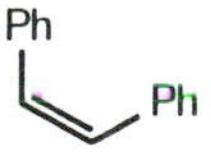

H3

Suggestions for further reading

An overview of the oxidation of alkenes can be found in Chapter 5 of F. A. Carey and R. J. Sundberg, *Advanced Organic Chemistry*, 3rd edition, Part B, Plenum, 1990, and 'directed' oxidation is reviewed in D. A. Evans, *Chem. Rev.*, 1993, **93**, 1307–1370. Chapter 3.2 of *Comprehensive Organic Synthesis*, (ed. B. M. Trost, I. Fleming, and S. V. Ley), Vol. 7, Pergamon, Oxford, 1991, details Sharpless epoxidation. Asymmetric dihydroxylation is reviewed in detail in H. C. Kolb, M. S. VanNieuwenhze, and K. B. Sharpless, *Chem. Rev.*, 1994, **94**, 2483–2547.

9 Rearrangements

9.1 General considerations

The most important types of rearrangement reactions which find use in stereoselective organic synthesis are 2,3- and 3,3-sigmatropic rearrangements, and the isomerization of alkenes. The sigmatropic rearrangements which we will cover are the Claisen rearrangement and its variants, and the 2,3-Wittig rearrangement. These rearrangements have been used for some time in organic synthesis, unlike the isomerization of alkenes. In this context, the latter has become important relatively recently owing to the discovery of very effective enantioselective catalysts.

9.2 Sigmatropic rearrangements

The Claisen rearrangement of allyl vinyl ethers (and related compounds) is probably the most widely used 3,3-sigmatropic rearrangement in stereoselective organic synthesis. The transition state for this type of reaction has been discussed in Chapter 2, and is closely analogous to that of the aldol reaction (Chapter 5). In many cases, analysis of the transition state as a chair conformation (Fig. 9.1) is appropriate, and will allow stereochemical predictions and correlations to be made.

Fig. 9.1 Chair transition state and chirality transfer in a Claisen rearrangement

Chiral substrates such as **9.1** (Fig. 9.1) generally undergo rearrangement through a chair transition state in which the large group attached to the stereocentre is equatorial. The double bond substituent (a in **9.1**) is also equatorial. The absolute stereochemistry and the double bond geometry in the product should then be as shown in Fig 9.1. The original stereocentre has been destroyed, and a new one created, and this type of process is often referred to as 'chirality transfer'.

This analysis can be generalized as shown in Fig. 9.2. In effect the stereochemistry of the product is determined by the chirality of the original stereocentre and the geometry of both double bonds.

The Claisen rearrangement and some of the important variants from the point of view of stereoselective organic synthesis are shown in Fig. 9.3. All involve the generation of an allyl vinyl ether or an analogous system. In all

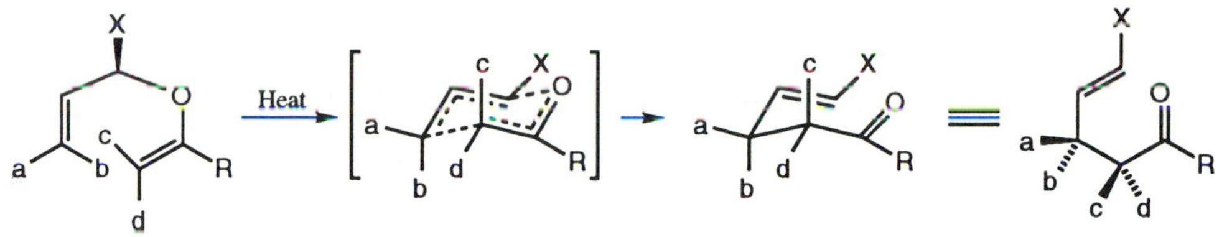

Fig. 9.2 Stereochemical correlations in Claisen rearrangements

but the Ireland/Claisen rearrangement, formation of the vinyl ether unit in highly geometrical purity is usually difficult to achieve. Accordingly the Claisen, Johnson/Claisen, and Eschenmoser/Claisen rearrangements tend to be used in cases where this is not an issue, i.e. in cases where substituents c and d are both hydrogen (Fig. 9.2).

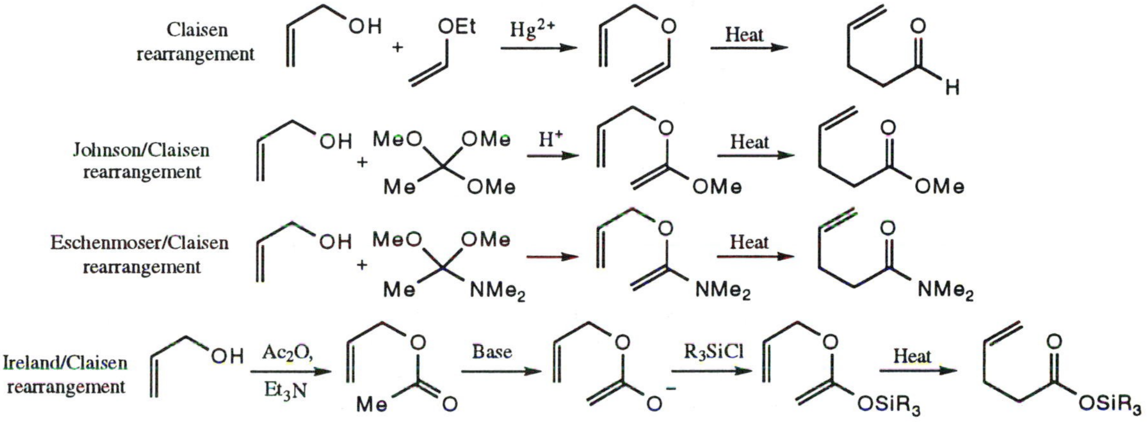

Fig. 9.3 Claisen rearrangement and variants

The Claisen rearrangement itself usually requires rather high temperatures, but can still be very effective. In the rearrangement of **9.2** (Fig. 9.4) the allyl ether is conformationally mobile, and the product corresponds to rearrangement in the conformation in which $A^{(1,3)}$ strain is minimized (i.e. with the C–H* bond more or less coplanar with the allylic double bond), and with essentially complete chirality transfer.

The level of stereochemical control which can be achieved is often very high, and can be put to good use in organic synthesis. If the alcohol **9.3** (Fig. 9.5) is resolved, and the alkyne group in each enantiomer is reduced as outlined, the resulting allylic alcohols **9.4** and **9.5** will give the same

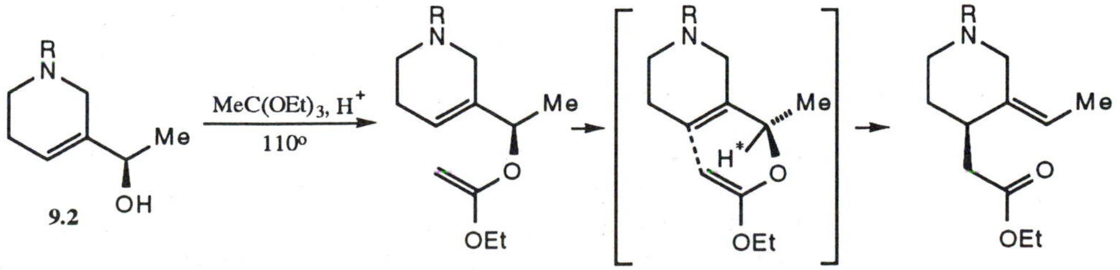

Fig. 9.4 Chirality transfer in Johnson/Claisen rearrangement

enantiomer of the product of Eschenmoser/Claisen rearrangement. In this way, both enantiomers of our starting material have been converted into the same enantiomer of the product. Such a process is referred to as 'enantioconvergent', and can be used to make resolution a relatively efficient process.

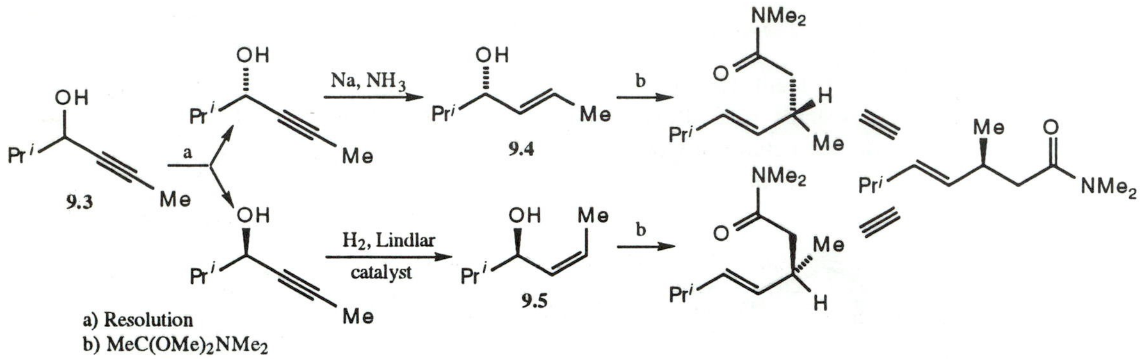

a) Resolution
b) $MeC(OMe)_2NMe_2$

Fig. 9.5 An enantioconvergent strategy

The Ireland/Claisen rearrangement, exemplified by the reactions of ester **9.6** (Fig. 9.6), usually involves formation of an enolate (**9.7** or **9.8**) which is trapped as the silyl ether (**9.9** or **9.10**). This provides the vinyl ether for the rearrangement. The enolate geometry determines the geometry of the vinyl ether, so control in the enolization step should result in control of the stereochemistry of the final product. In practice it is often possible to control enolate geometry simply by choice of enolization conditions. Use of THF alone as solvent usually produces the (*Z*) enolate, whereas in the presence of the dipolar aprotic solvent HMPA the (*E*) enolate tends to predominate (Fig. 9.6). The consequence of this is that either diastereoisomeric product **9.11** or **9.12** can be formed simply by choice of enolization conditions.

HMPA = $[(Me_2N)_3P{=}O]$

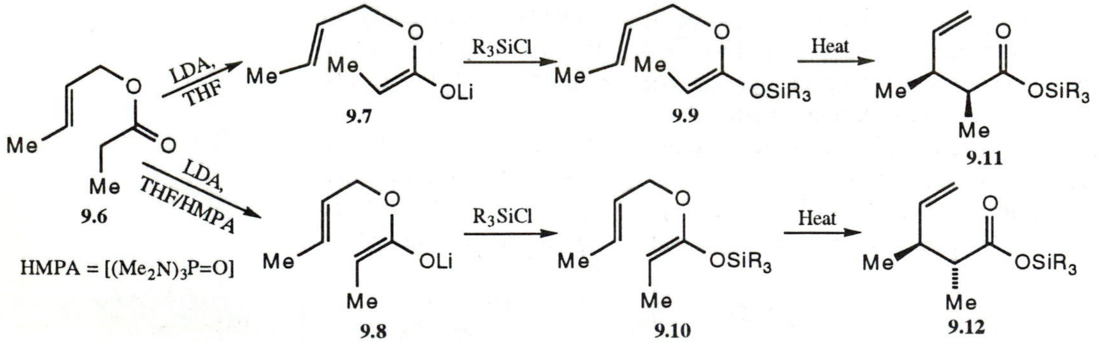

Fig. 9.6 Specific enolates lead to specific stereoisomeric products in an Ireland/Claisen rearrangement

It is possible to carry out highly enantioselective Claisen rearrangements of prochiral esters using the chiral boron reagent **9.13** (Fig. 9.7). This reagent has also been used in aldol reactions (Chapter 5). The geometry of enolization again depends on the reaction conditions, and the resulting boron

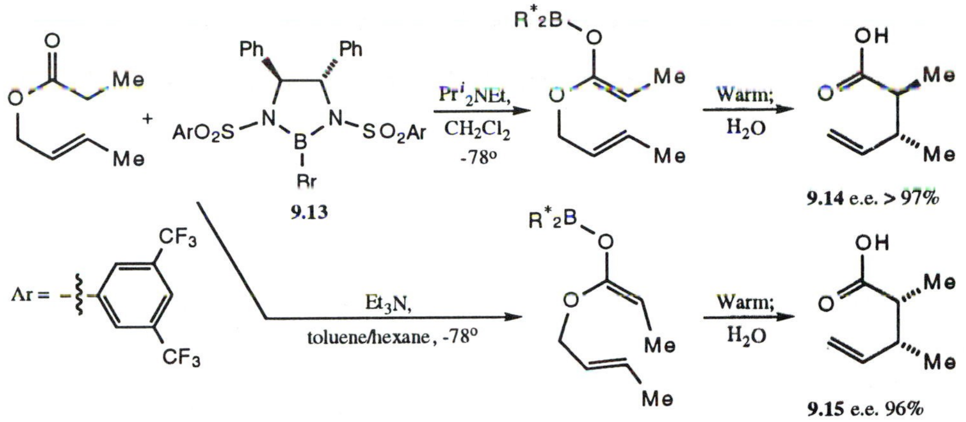

Fig. 9.7 Enantioselective Claisen rearrangement

enolates undergo rearrangement to give either of the acids **9.14** or **9.15** in high enantiomeric excess.

The other type of sigmatropic rearrangement which we will cover is the 2,3-Wittig rearrangement, as outlined in Fig. 9.8. Usually, the substrate **9.16** contains a group X which stabilizes the anion **9.17**, but this is not necessary if the anion is formed by an alternative method (see Fig. 9.9). The rearrangement is driven by the formation of the relatively stable alkoxide **9.18**, and completed by protonation in the reaction work up.

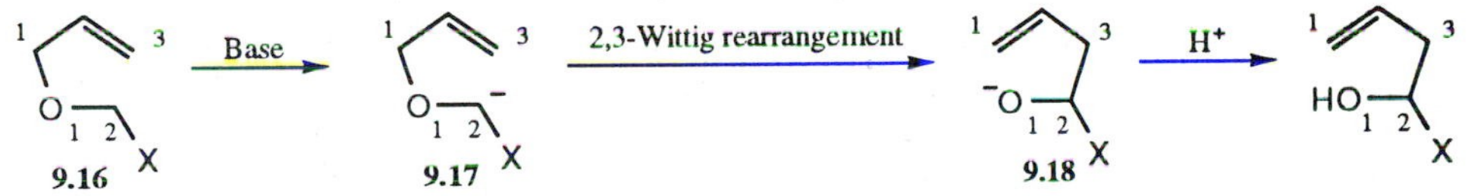

Fig. 9.8 The 2,3-Wittig rearrangement

In the absence of an anion stabilizing group X, the required carbanion can be generated from the trialkylstannane by a lithium-tin exchange process, as in the rearrangement of **9.19** (Fig. 9.9). This reaction also shows that as with 3,3-sigmatropic rearrangements, there can be a high level of chirality transfer in the 2,3-Wittig rearrangement.

Usually an anion stabilizing group is present, and stereochemical issues then arise. The stereochemistry of the allyl ether double bond is important, as in the rearrangement of **9.20** (Fig. 9.10). By analogy with 3,3-sigmatropic rearrangements, with a stereocentre present in the substrate (e.g. **9.21**)

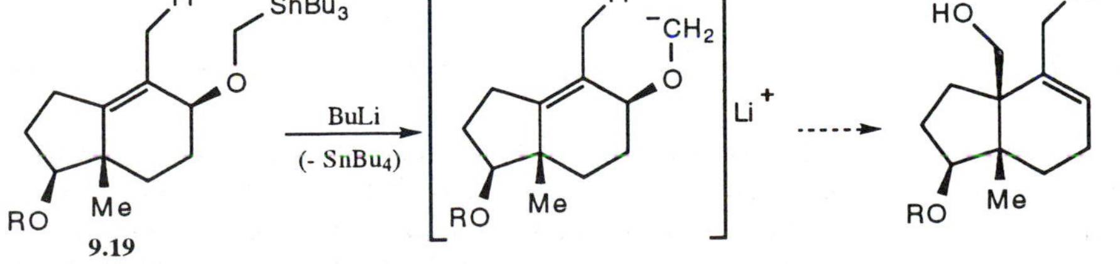

Fig. 9.9 2,3-Wittig rearrangement in the absence of an anion stabilizing group

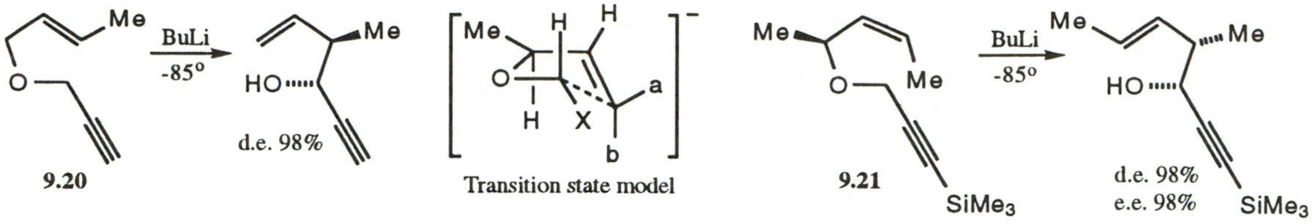

Fig. 9.10 Stereochemical relationships in 2,3-Wittig rearrangements

chirality transfer can be high. There are similarities to 3,3-sigmatropic rearrangements in the correlation of substrate stereochemistry (and geometry) with product geometry and stereochemistry. In contrast, there is some debate as to the most appropriate transition state model, but one of the most useful models is illustrated for the reactions of **9.20** and **9.21** (Fig. 9.10).

As for the Claisen rearrangement, it is possible to carry out enantioselective 2,3-Wittig rearrangements with an appropriate chiral boron derivative. A version of that used for the Claisen rearrangement has been reported to give high enantiomeric excess in the 2,3-Wittig rearrangement of substrates such as **9.22** (Fig. 9.11). The proposed transition state involves co-ordination to the central boron to give a tetrahedral boron chelate, and rearrangement so that steric interactions are minimized.

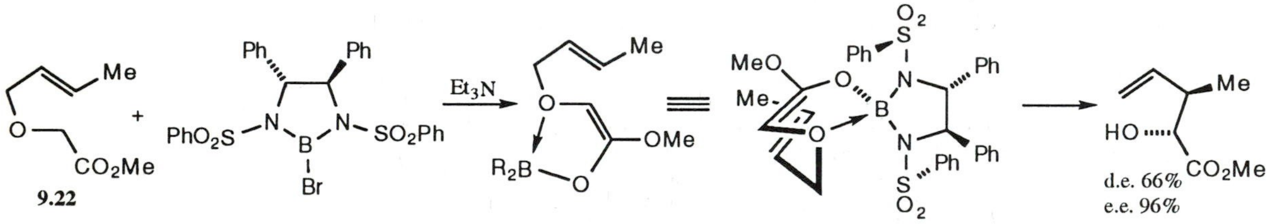

Fig. 9.11 Enantioselective 2,3-Wittig rearrangement

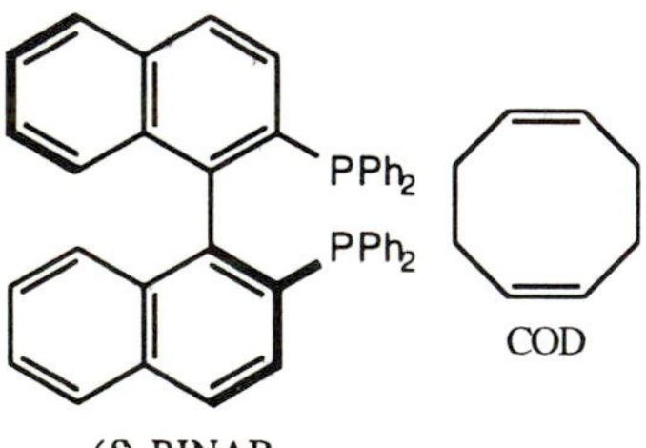

9.3 Alkene isomerization

Rearrangement of alkenes by a 1,3 shift of an allylic hydrogen has become important as a result of the development and application of efficient enantioselective catalysts. The best substrates are prochiral allylic amines which can rearrange into chiral enamines (Fig. 9.12). The catalysts which are used are closely related to those used for enantioselective homogeneous hydrogenation, and use either enantiomer of BINAP as the chiral ligand.

Mechanistic studies have revealed that the reaction is probably a concerted suprafacial 1,3-hydrogen shift, via a square planar transition state. As would be expected for such a process, chirality transfer is very high and the geometry of the double bond of the allylic amine will determine the absolute configuration of the enamine for a given enantiomer of BINAP (Fig. 9.13).

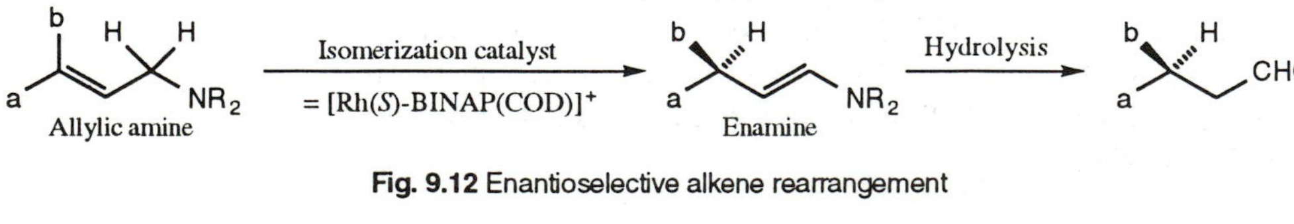

(*S*)-BINAP

COD

Fig. 9.12 Enantioselective alkene rearrangement

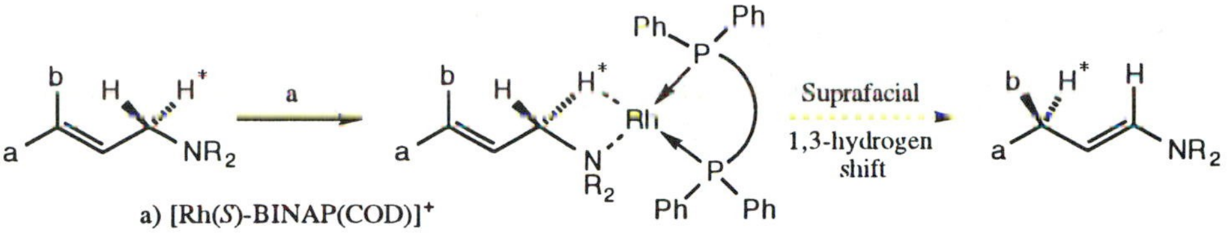

a) [Rh(*S*)-BINAP(COD)]$^+$

Fig. 9.13 Mechanism of enantioselective alkene rearrangement

This type of enantioselective isomerization occupies a key place in a commercial synthetic process (Fig. 9.14) which produces greater than 1500 Tonnes per annum. The so called 'Takasago Process' provides a significant proportion of the annual world production of (–)-menthol.

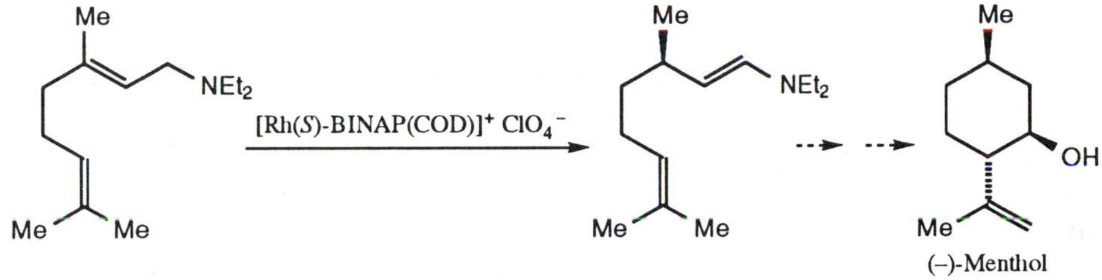

Fig. 9.14 Key enantioselective catalytic step of the Takasago Process

9.4 Problems

9.1 Eschenmoser/Claisen rearrangement using **J1** gives allylsilane **J2**. Which enantiomer of **J2** will be produced?

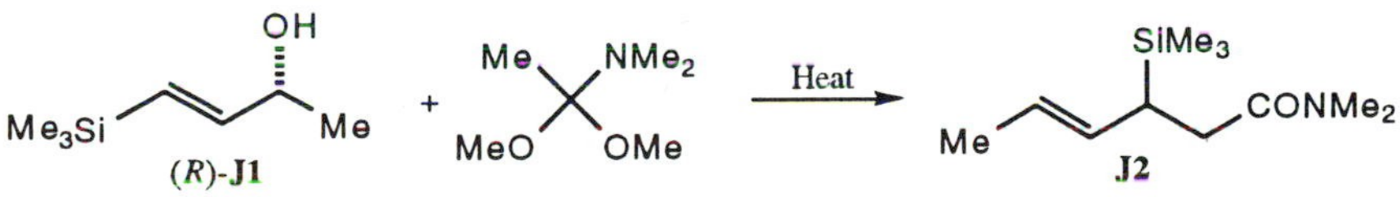

9.2 Which diastereoisomer will predominate in the rearrangement of **J3**?

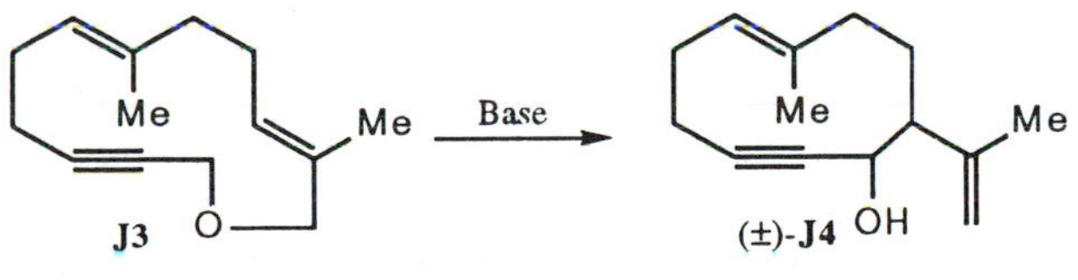

Suggestions for further reading

The Claisen and 2,3-Wittig rearrangements are discussed in Chapter 6 of F. A. Carey and R. J. Sundberg, *Advanced Organic Chemistry*, 3rd edition, Part B, Plenum, 1990, and the stereochemical aspects are detailed in Chapter 12 of E. L. Eliel, S. H. Wilen, and L. N. Mander, *Stereochemistry of Organic Compounds*, Wiley, New York, 1994. For an introduction to the enantioselective rearrangement of allylic amines, see S. Otsuka and K. Tani, *Synthesis*, 1992, 665–680.

10 Hydrolysis and esterification

10.1 General considerations

The reactions with which we will be concerned in this chapter are enzyme catalysed hydrolysis of esters and esterification. From our point of view, we can regard enzymes simply as enantioselective catalysts. They are much larger and much more complicated than 'normal' enantioselective catalysts, so a detailed understanding and prediction of both the sense and level of enantioselectivity is often rather difficult. In addition, many enzymes are selective about the structure of the substrates which they will accept. Nevertheless, there are some enzymes which are cheap, available in quantity, and which will accept a reasonably wide range of substrates. In general it is these which find significant application in stereoselective synthesis.

Enzyme catalysed kinetic resolutions are essentially a technique for racemate resolution, rather than stereoselective synthesis, and so will not be considered here.

There are two basic processes which can be carried out by enzyme hydrolysis, resolution of a racemate by kinetic resolution, and conversion of a *meso*-compound into a chiral product (Fig. 10.1). Idealized general examples of these processes are represented in Fig. 10.1.

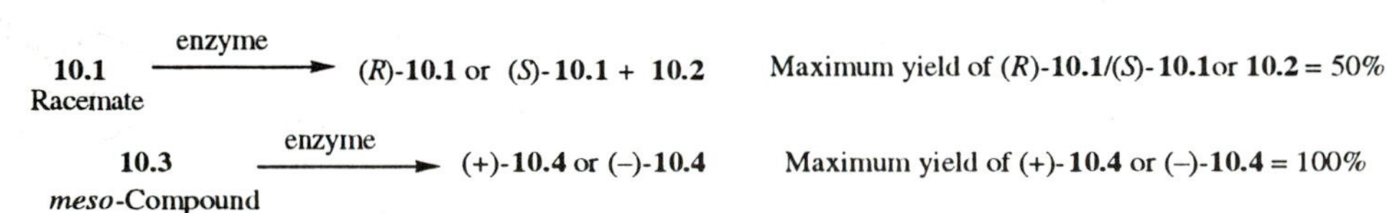

Fig. 10.1 Comparison of resolution and 'asymmetrization'

10.2 Hydrolysis/esterification of *meso* compounds

Most of the reactions in this area involve ester hydrolysis, and one of the most widely used enzymes is pig liver esterase. Typically, the *meso*-compound used as starting material possesses two enantiotopic ester groups (CO_2Me_A and CO_2Me_B, Fig. 10.2), so hydrolysis of one of these will produce enantiomeric products. This is sometimes referred to as 'asymmetrization' (Fig. 10.2).

Careful chemical manipulation of a single enantiomer produced by asymmetrization can give products in either enantiomeric form. This is often referred to as the '*meso*-trick', and a typical example is shown in Fig. 10.3,

CO_2Me_A / CO_2Me_B —Pig liver esterase→ CO_2H / CO_2Me_B **10.5** e.e. 96%

CO_2Me_A / CO_2Me_B —Pig liver esterase→ CO_2H / CO_2Me_B e.e. > 97%

Fig. 10.2 Asymmetrization of *meso*-diesters with pig liver esterase

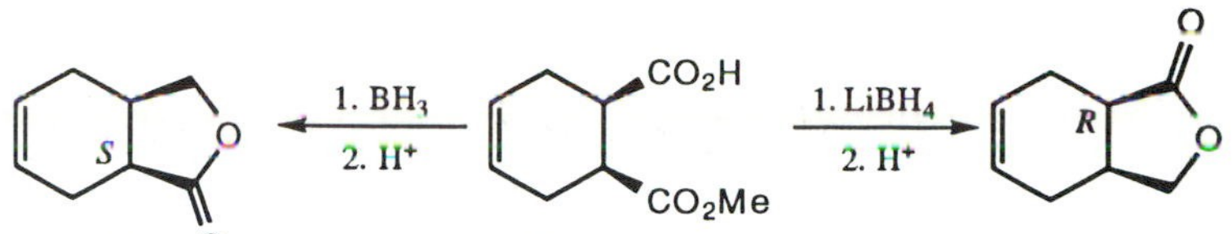

Fig. 10.3 The '*meso*-trick'

where a single enantiomer of **10.5** is used to produce both enantiomers of the lactone **10.6** by selective manipulation of the functional groups.

The origin of the enantioselectivity of enzyme catalysed ester hydrolysis lies in the fact that enzymes are chiral, and present as a single enantiomer. They are proteins built up from a relatively large number of amino acids (which are usually (*S*)-amino acids). Many of the esterases possess a serine residue at the active site (Fig. 10.4), which reacts with the ester of the substrate, to form an acylated enzyme intermediate. This is subsequently hydrolysed, releasing the acid and the enzyme for another catalytic cycle.

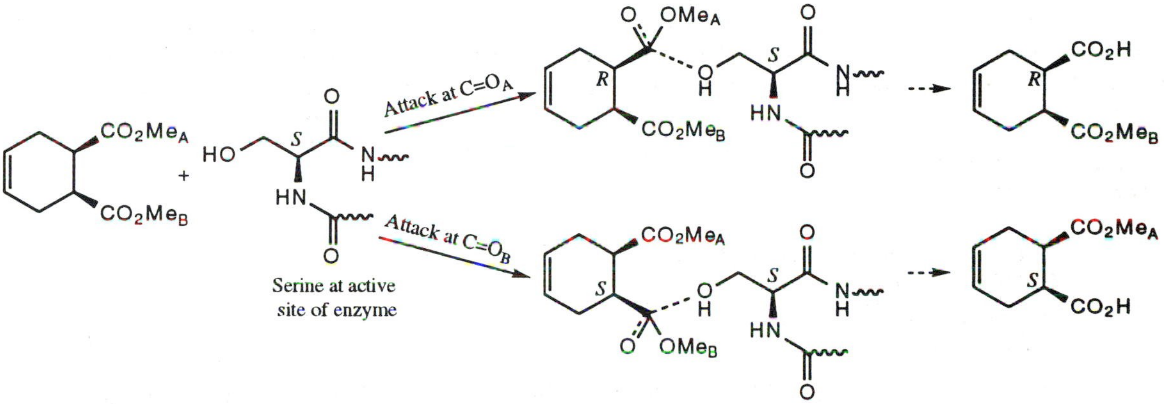

Fig. 10.4 Diastereoisomeric transition states in asymmetrization of a *meso*-diester

The ester groups of the *meso*-compound are enantiotopic, and the enzyme is chiral, so reaction of the serine with the esters will produce diastereoisomeric intermediates, via diastereoisomeric transition states (Fig. 10.4). Thus the enzyme is capable of distinguishing the enantiotopic groups.

It is also possible to produce chiral alcohols in high enantiomeric excess, rather than chiral acids. Hydrolysis of the *meso*-cyclopentane diacetate **10.7** catalysed by acetyl cholinesterase which provides the monoacetate **10.8** by selective hydrolysis of ester 'A' in **10.7**. Hydrolysis of the closely related diacetate **10.9** with the same enzyme also gives the monoacetate (**10.10**) in high enantiomeric excess, but hydrolyses other ester group ('B')!

Ac = CH_3CO

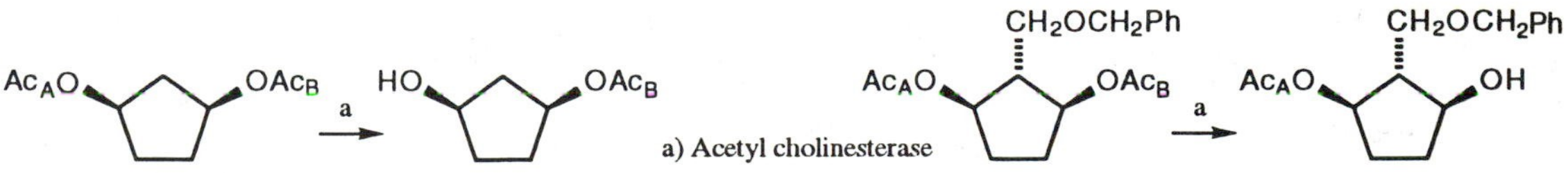

Fig. 10.5 Asymmetrization of esters of *meso*-diols

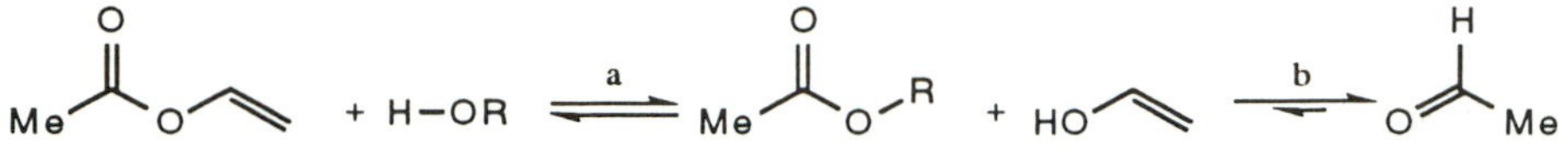

10.11 Vinyl acetate + H−OR ⇌(a) **10.13** + HO(**10.14**) ⇌(b) **10.15**

10.11 Vinyl acetate **10.12** **10.13** **10.14** **10.15**

a) Enzyme catalysed transesterification b) Keto-enol tautomerism

Fig. 10.6 Transesterification using vinyl acetate

Esterase enzymes can also be used to catalyse the reaction of an ester with an alcohol to produce a new ester, a process known as transesterification. This process is reversible, which can lead to mixtures. However, if vinyl acetate (**10.11**) is used as the ester, transesterification will produce a mixture of the new ester (**10.13**) and vinyl alcohol (**10.14**, Fig. 10.6). Vinyl alcohol is the enol tautomer of acetaldehyde (**10.15**), and the second equilibrium lies far to the right. As vinyl alcohol is produced, it converts to acetaldehyde, and the transesterification equilibrium is driven in favour of the new ester.

This transesterification approach can be very effective, and has been used in a synthesis of the natural enantiomer of aristeromycin (Fig. 10.7). The *meso*-diol **10.16** undergoes transesterification with vinyl acetate to produce the monoacetate **10.17** in high enantiomeric excess, which was subsequently converted via a number of steps to aristeromycin.

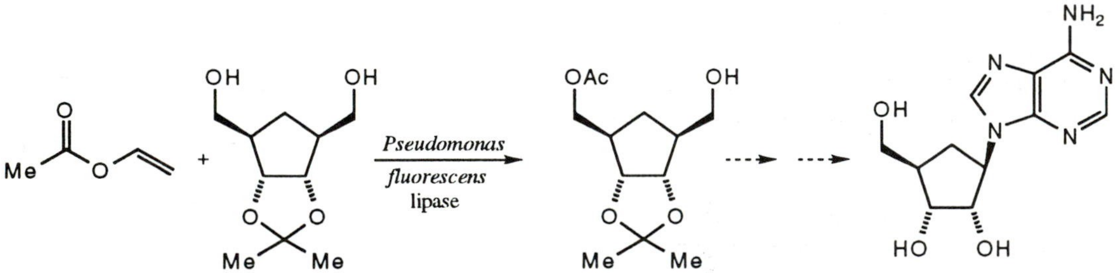

Pseudomonas fluorescens lipase

10.16 **10.17** e.e. > 99% Aristeromycin

Fig. 10.7 Asymmetrization using enzyme catalysed transesterification with vinyl acetate

Suggestions for further reading

For an introduction to enzymes as enantioselective catalysts see Chapter 5 of *Asymmetric Synthesis*, (ed. J. D. Morrison), Vol. 5, Academic Press, New York, 1985. An outline of the scope of enzyme hydrolysis using pig liver esterase is provided in L.–M. Zhu and M.C. Tedford, *Tetrahedron*, 1990, 46, 6587–6611, and enzyme catalysed transesterification is detailed in K. Faber and S. Riva, Synthesis, 1992, 895–910.

Answers to problems

1.1 The product would be expected to have an e.e >0% (see Chapter 10).

1.2 (i) Stereospecific; (ii) Stereospecific; (iii) Stereoselective.

2.1 **B1** Gives *meso*-2,3-dibromobutane. **B2** Gives alcohol (±)-**B7**.

2.2 In **B3** and **B4** the phenyl group is equatorial, so the alignment of the alkoxide from **B3** is fine for an internal S_N2 reaction to form **B5**. The alkoxide from **B4** has a C–H bond antiperiplanar with the leaving group, so this bond migrates (**B8**), forming the enolate of **B6**.

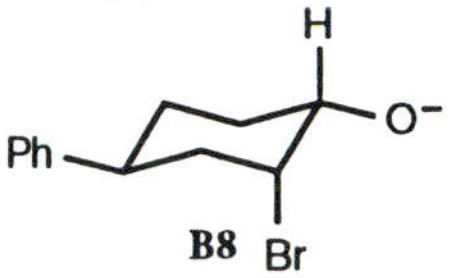

3.1 The major product should be **C2**.

3.2 The major product should be **C6**.

3.3 The major product should be **C9**.

4.1 **D3** Should predominate.

4.2 The enolate adopts a conformation which minimizes $A^{(1,3)}$ strain, and alkylates from the face opposite to the bulky Ph_3Si group.

5.1 The product **E2** should be the *anti*-aldol.

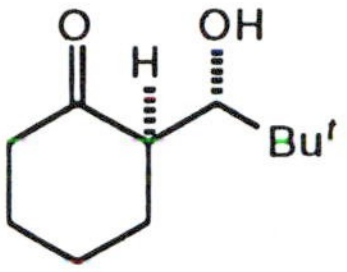

5.2 The maximum co-ordination number of boron is 4. The aldehyde carbonyl 'displaces' the carbonyl of the oxazolidinone, which is now free to rotate and adopt a conformation which minimizes dipole-dipole repulsions due to its now 'free' carbonyl group.

6.1 The major product should be **F2**, assuming that $A^{(1,3)}$ strain 'controls' the conformation of the alkene, and that the hydroboration takes place *anti* to the $PhMe_2Si$ group.

6.2 A chelated titanium enolate is formed, and the electrophilic α,β-unsaturated ketone approaches from the face of the enolate opposite to the CH_2Ph group.

6.3 The major product is expected to be **F7**.

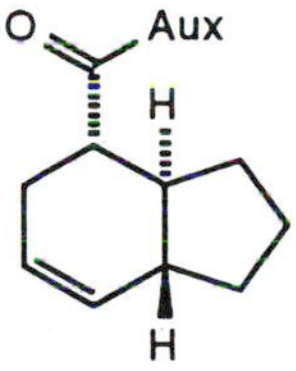

7.1 Both reactions should give (*S*)-citronellol.

7.2 Alcohol **G3** is the product of reduction in the conformation in which dipole–dipole repulsions are minimized. Alcohol **G4** is the product of 'chelation controlled' reduction.

7.3 Assuming that the reduction follows the trends illustrated in Fig. 7.17, reduction of one of the C=O groups in **G3** should give two diastereoisomeric β-hydroxy ketones **G7** and **G8**. *Both* of these give the same diol **G6** on reduction of the C=O group (using the same model).

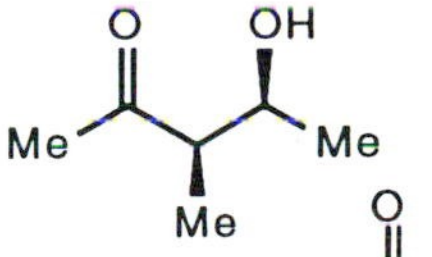

8.1 **H1** requires a D-(–)-tartrate diester, and **H2** a L-(+)-tartrate.

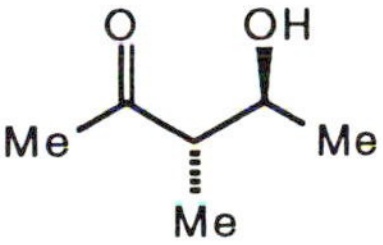

8.2 No, as both reactions would give *meso*-products, using an enantioselective oxidation method is pointless.

9.1 The allylsilane should possess the (*R*)-configuration.

9.2 The product of 2,3-Wittig rearrangement, using the model illustrated in Fig. 9.10, should be (±)-**J4**.

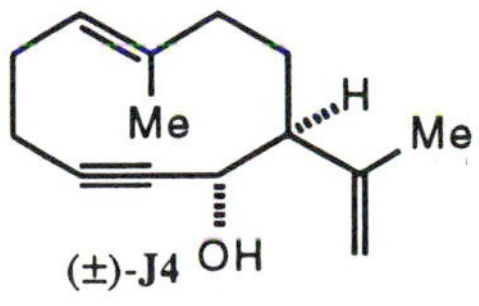

(±)-**B7**

B8

(±)-**E2**

F7

G7

G8

(±)-**J4**

Index